ECONOMIC COMMISSION FOR EUROPE

GENEVA

ENERGY EFFICIENT DESIGN

A guide to energy efficiency and solar applications in building design

ECE ENERGY SERIES No. 9

UNITED NATIONS

NEW YORK, 1991

NOTE

Symbols of United Nations documents are composed of capital letters combined with figures. Mention of such a symbol indicates a reference to a United Nations document.

*

* *

The designations employed and the presentation of the material in this publication do not imply the expression of any opinion whatsoever on the part of the Secretariat of the United Nations concerning the legal status of any country, territory, city or area, or of its authorities, or concerning the delimitation of its frontiers or boundaries.

Text: Iain Borden, Adrian Leaman and Mariana Atkins
Graphics: Carmel Lewin and Iain Borden

UNITED NATIONS PUBLICATION
Sales No. E.91.II.E.31
ISBN 92-1-116519-9
ISSN 1014-7225

contents

Contents

preface

Solar Applications and Energy Efficiency in Building Design and Town Planning (RER/87/006) is a United Nations Development Programme (UNDP) project of the Governments of Albania, Bulgaria, Cyprus, The Czech and Slovak Federal Republic, France, Hungary, Malta, Poland, Turkey, United Kingdom and Yugoslavia. The project began in 1988 and comes to a conclusion at the end of 1991. It is to enhance the professional skills of practising architects, engineers and town planners in European IPF* countries to design energy efficient buildings which reduce energy consumption and make greater use of passive solar heating and natural cooling techniques.

The United Nations Economic Commission for Europe (ECE) is the Executing Agency of the project which is implemented under the auspices of the Committee on Energy, General Energy Programme of Work for 1990-1994, sub-programme 5 Energy Conservation and Efficiency (ECE/ENERGY/15). The project has five main outputs or results: an international network of institutions for low energy building design; a state-of-the-art survey of energy use in the built environment of European IPF countries; a simple computer program for energy efficient building design; a design guide and computer program operators' manual; and a series of international training courses in participating European IPF countries.

Energy Efficient Design is the fourth output of the project. It comprises the design guide for practising architects and engineers, for use mainly in mid-career training courses, and the operators' manual for the project's computer program. The design guide and computer program were prepared by the project's sub-contractor Building Use Studies Ltd., London (England) during 1990 under United Nations contract no. G/CON/01/UN in consultation with experts in participating countries. Enquiries about the project should be addressed to

The Energy Division,
Economic Commission of Europe,
Palais des Nations,
CH 1211 Geneva 10,
Switzerland

Telephone 734-6011
Telex 412-692 UNO CH
Telefax 733-9879

*Indicative Planning Figure (IPF) countries are those entitled to receive United Nations technical assistance.

v

introduction

Energy Efficient Design is an educational aid which provides the knowledge framework necessary for the design of buildings from the perspective of energy efficiency. It can be used as a stand-alone tool, or in conjunction with the companion computer software and training programme.

The process of designing buildings has five stages: understanding the need, defining the brief, design activity, implementing the design, and monitoring buildings in use. *Energy Efficient Design* is above all a guide to the first two of these stages which helps designers to understand the need for energy efficiency and to define a brief which keeps energy efficiency in mind by explaining the basic characteristics and inter-relations of the different factors involved. The guide also refers to the fifth design stage and the different ways of monitoring buildings. In the interests of comprehension, specialized terminology is avoided wherever possible.

This guide does not tell people how to design or what an energy-efficient building should look like; the emphasis is on principles, not details, of design. Because

conditions vary greatly between countries and over time, they are best understood by those people who are closest to them. Responsibility therefore rests with designers to make the guide work in practice, as befits their own circumstances. This approach - combining design guidance with local skills and knowledge - ensures that all available resources are fully exploited and is essential if the lessons learned are to be applicable across the broadest possible range of countries, climates and buildings.

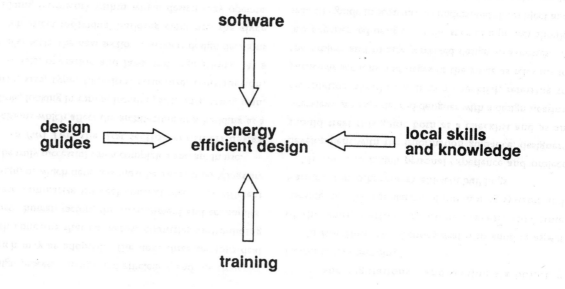

Above

The resource base for the energy-efficient design of buildings. The same kinds of information may be supplied by more than one of these resources. For example, technical information may be supplied by a design guide, by a piece of computer software or by local building professionals.

All resources must be properly co-ordinated if energy-efficient design is to be successful. An understanding of the management requirements and capabilities is therefore also required.

using the guide

Energy Efficient Design allows designers to focus attention at a number of levels, ranging from the design process itself and global-level concerns such as energy resources to building-level concerns such as the design of floors, walls and windows.

Each part of the guide corresponds to a particular stage of the design process: *Part One Understanding Energy Efficiency* provides an introduction and overview, giving designers a basic understanding of the main themes and concerns; *Part Two Defining the Brief* identifies the various factors relevant to the energy-efficient design process, telling designers what they need to consider if buildings are to be energy efficient; *Part Three Monitoring* explains the importance of having reliable information about how buildings actually perform in use; and *Part Four* considers the advantages of using computer software for energy-efficient design, and explains the UNDP-ECE program which accompanies this guide.

Within *Parts One, Two and Three*, sections are again arranged hierarchically. In *Part One*, the first two sections deal with fundamental aspects of the energy-efficient

design process: energy and efficiency, and the strategies which may be adopted. The next three sections deal with concerns that cut across countries and building types: human factors, the environment and economics. These summarize the real context of energy-efficient design, of which designers must be aware if buildings are to be truly successful when completed and are in use.

In *Part Two*, the first section considers design decisions which affect the architecture of a building as a whole, looking in turn at location, site, orientation, plan, form, size, type, function, structure, construction, materials, operation and laws and regulations. At a smaller scale, the next section considers design decisions which affect individual building elements; the given building framework within which design may operate. The last section describes specialized solar applications designed with energy efficiency in mind.

Urban design issues may be found both in Part One and in Part Two. In particular, see section four Environment, section five Architecture "Location", "Site", "Construction and material", "Operation" and

"Laws and regulations", and section six Building Elements "Landscaping".

In *Part Three*, two sections deal with another aspect of the energy-efficient design process, this time considering the installation of monitoring systems and learning from other energy-efficient buildings.

In order to match personal experience and project development with the information guidance, designers should treat this guide both as a checklist and as an overview. An experienced designer with a design nearing completion should use it as a checklist, referring to particular sections and pages of the guide as relevant to the project and to any perceived design weaknesses. A less experienced designer at the start of a project should read the guide in sequence to understand the subject and to develop a list of requirements for the design in question. In either case, careful comparison of the guide with the building design will ensure that all relevant factors have been considered. Eventual success remains dependent on the skill of the designer in dealing with the restrictions and opportunities at hand.

part one

understanding energy efficiency

what is energy?

Energy is the capacity for doing work, and is ultimately degraded into heat. The use of fuels, rather than muscle power and ambient sources, for energy has been an important determining factor in history. There are six main sources of energy: fossil fuels such as coal, oil and natural gas; wind and water; nuclear power; wood and other biological materials; geo-thermal power; and solar radiation. Quantitatively, solar radiation is the largest energy source. Some sources, such as coal and oil, are abundant in certain locations and have the potential to be used anywhere. Unlike local sources such as firewood and dung, they offer high energy content, easy storage and transportation, and flexibility of application.

Differences in energy use across countries depend on regional and national circumstances. For example, economic power and technical knowledge determine both access to energy resources and the potential to convert them into a usable form. Countries have moved from ad hoc local resources to processed and transported fuels, but they run the risk of becoming overly dependent on certain energy types. This makes national economies susceptible to changes in supply conditions, as has occurred with oil in the 1970s and 1990s. Developing countries are especially vulnerable in this respect.

Looking ahead, factors of increased demands, the finite nature of many sources, environmental problems and the burden placed on national economies make dependence on processed fuels less viable. There is therefore a need to use available and underdeveloped resources in a more efficient manner. The energy-efficient design of buildings is one response to this need.

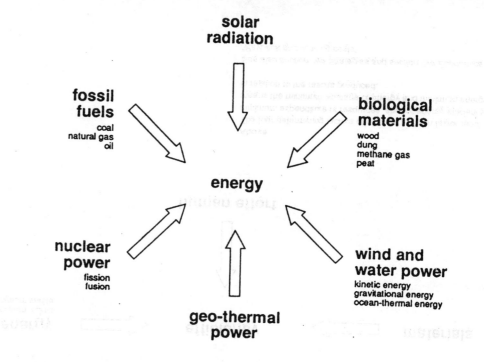

solar
radiation

fossil
fuels
coal
natural gas
oil

biological
materials
wood
dung
methane gas
peat

energy

nuclear
power
fission
fusion

wind and
water power
kinetic energy
gravitational energy
ocean-thermal energy

geo-thermal
power

Above
The six main sources of energy.

See also section two Strategies "Controlling gains" and section four Environment "Environmental concerns".

what is energy efficiency?

Efficiency is the capacity to produce results with a minimum overall expenditure of energy, human effort, materials and capital. Because the cost and availability of these resources differ, efficiency is relative to time and place. This means that countries have different perspectives and limitations on efficiency, but share the same aim of minimizing expenditure of resources for the task at hand.

Energy efficiency is the capacity to produce results with a minimum expenditure of energy inputs. Waste energy is any energy in a system which is not directly serving the system's functions. A system is thought of as energy-efficient if its requirements are low in relation to the results produced, and if energy is not wasted.

For example, programmes of insulation in dwellings have been aimed at lowering the quantity of energy required to overcome heat losses through walls and roofs. In order to be energy efficient, these programmes must be matched by an effective heating system with high conversion efficiency, minimum distribution losses and good controls.

capital

energy
minimum inputs
minimum waste

efficiency

materials

human effort

Above
The four determining factors of efficiency. Each factor requires the minimum expenditure of resources, and for energy efficiency this means the minimum wastage of energy and minimum energy inputs in relation to the results produced.

See also section two Strategies and section five Economics "Payback and running costs".

what is energy efficient design?

All buildings require energy inputs at various points in their life cycle. The amount, type and uses of energy vary across buildings, regions and countries. In general, combustible fuels and electricity are supplied for lighting and equipment, cooking, hot water, heating and cooling, and mechanized access. Other energy inputs are derived as incidental heat gains from these services, from the sun and from people. Solar radiation in particular can make significant contributions to heating, cooling and lighting loads. In "low energy" buildings designed to minimize energy consumption, these incidental gains can meet 30-60 per cent of the total heat requirement.

Energy-efficient design embraces the design and building process, the building as it is intended to be used, and the building as it is actually used. Energy-efficient design therefore prioritizes energy efficiency for human factors, the environment, economics, architecture, building elements, solar applications and monitoring.

The design product, the building, is then energy efficient, although some buildings may be energy efficient without having been consciously designed from this perspective. An energy-efficient building will have the minimum levels of energy input and energy wastage compatible with its function, while meeting all performance standards for aesthetics, health, safety and use.

Energy-efficient design demands an exchange of energy-conscious information among designers, building managers and users. This exchange can occur at two levels: short-term feedback at local level, and long-term feedback at global level.

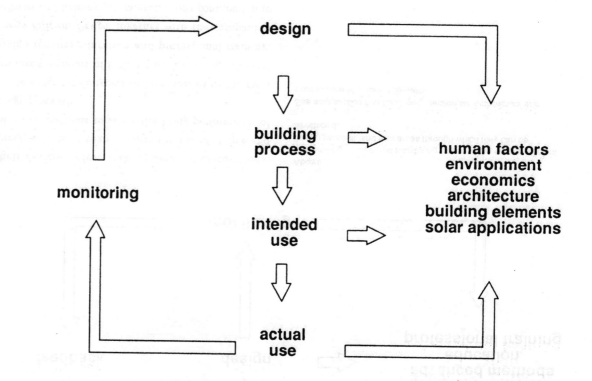

Above
The process of energy-efficient design.

See also section six Architecture, section seven Building Elements, section eight Solar Applications and section nine Monitoring Systems.

method

Energy-efficient design is sensitive to how buildings perform as environmental systems and to how buildings are used. Therefore, the briefing, design and monitoring stages of the design process are of equal importance. Rather than concentrating on any one of these stages, it is the systematic implementation of all three which characterizes energy-efficient design.

Briefing and monitoring. The relationship between brief and performance in use is critical to energy-efficient design. An energy-efficient building normally has a brief which defines space use, layout and performance standards, and which is detailed enough for design criteria to be compared with actual performance later on. Experimental test buildings are often established with this intention. Detailed knowledge is first obtained by systematic monitoring, and is then passed back to designers.

Design. Buildings are complex systems with many different variables interacting with each other, making it difficult for architects and engineers to predict results. Designers must therefore understand the ramifications of

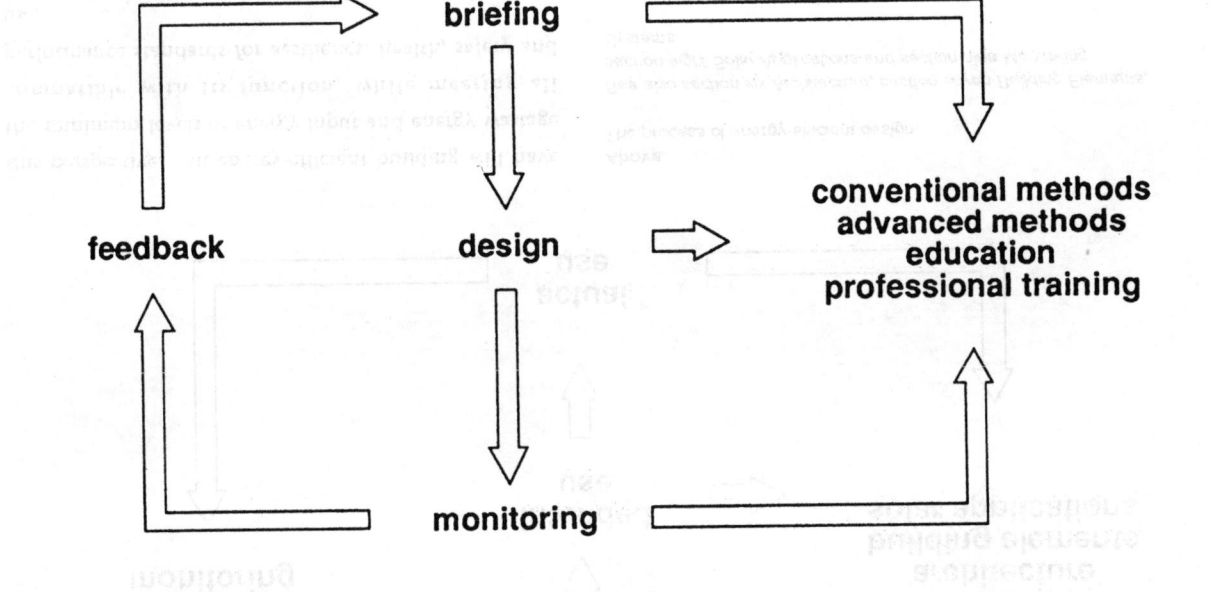

their designs before making detailed calculations. Energy-efficient design requires the use of simple and advanced methods to assess the likely performance of design proposals.

In order for designers to keep abreast of the most advanced methods of building design, energy-efficient design requires education and professional training. *Energy Efficient Design*, together with the companion software and training programme, is one contribution to this process.

Above
The briefing, design and monitoring stages of the energy-efficient design process, and four areas through which they can be developed.

See also section two Strategies, section six Architecture and section nine Monitoring Systems.

strategies

One of the main functions of a building is to create an environment which is less varied and therefore more usable than that outside. Energy-efficient design enables this system to be controlled from the perspective of energy efficiency: it gives buildings the ability to control energy gains, losses and demands, and gives users the ability to control the building and its technology. The first three of these are straightforward: controlling gains ensures best use is made of available energy inputs, preferably those that are renewable and involve minimum expenditure of resources; controlling losses ensures that energy outputs and wastage are minimized; and controlling demands ensures no more energy is asked for than is needed.

Energy efficiency is also affected by the characteristics of the services within the building. The technology selected for these services should be appropriate for context, cost, function and use, and should be capable of being installed, operated and maintained without undue difficulty.

In terms of an overall strategy, energy-efficient

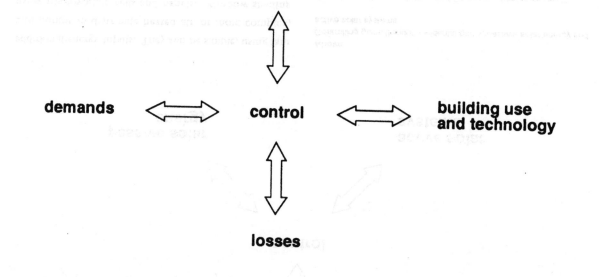

Above
Four strategies for energy-efficient design.

design also provides a feedback loop for designers to understand and act upon the consequences of design decisions, and to "design in" more control . An energy-efficient design strategy therefore embraces the whole building project, including design and construction process, intended use, and actual use, all within a responsive and functionally appropriate environment.

controlling gains

Control over energy gains in buildings may be exercised in three ways.

Incidental gains. Lights, equipment, cooking, hot water systems and people produce heat which can help reduce space heating requirements. However, these gains are not always useful, causing over-heating and energy demands for cooling. Energy-efficient design helps to maximize the efficiency of the systems and to minimize unwanted gains, particularly those which create additional demands for energy.

Passive solar energy. Passive solar design uses building form and fabric to admit, store, distribute and re-emit solar energy by radiation, conduction and natural convection. Passive systems obtain solar energy from glazing (direct gain), thermal mass (indirect gain from solar energy, first absorbed by the thermal mass and then radiated or convected as heat), or a separate storage system (isolated gain, usually heat transferred from a thermal mass away from the main building).

Active solar systems. Active solar systems work on similar principles to passive solar systems, but require additional energy inputs. They can be simple, using fans and pumps to distribute heated air, or more complex, using photovoltaic cells and panels. Window shading devices may be operated electrically to vary solar gain.

Any heating requirements remaining after the exploitation of available incidental and solar gains will require fuel energy inputs.

incidental
gains

control

passive solar
energy

active solar
systems

Above
Controlling gains through incidental gains, passive solar energy and active solar systems.

See also section one Energy and Efficiency "What is energy?", section six Architecture, section seven Building Elements and section eight Solar Applications.

controlling losses

Control over energy losses in buildings may be exercised in three ways.

Heat transfer and insulation. To reduce energy lost by the transfer of heat to the outside, the thermal properties of the building envelope (particularly thermal capacity and thermal transmittance) are modified. Materials are chosen in part for their thermal performance, and are used in energy-efficient configurations, such as the use of cavity walls or double-glazing in windows.

Insulation materials, whose thermal properties resist the transfer of heat, are incorporated into new designs or added retrospectively to existing buildings. Insulation is one of the most important and cost-effective methods of controlling heat losses.

Heat losses are also reduced by adjusting building location and orientation to the local climate, and by reducing the ratio of building perimeter to internal area.

Ventilation and infiltration. Heat losses may be reduced by closing windows, doors and other apertures and by sealing gaps around them. Gaps may also be sealed elsewhere in the structure, particularly at

junctions. Energy lost through apertures is often noticed only after a period of time. Control, feedback and user awareness are therefore critical. Ventilation may sometimes be beneficial in reducing or avoiding energy demands for cooling.

Efficiency. Energy is lost through inefficient systems and equipment, especially those for heating, cooling and lighting. All energy-consuming systems and appliances in buildings should have the lowest possible inputs and levels of wastage compatible with their function.

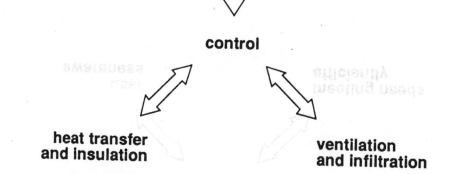

efficiency

control

heat transfer and insulation

ventilation and infiltration

Above
Controlling losses through heat transfer and insulation, ventilation and infiltration, and efficiency.

See also section three Human Factors "Control", section six Architecture, section seven Building Elements, and section eight Solar Applications.

controlling demands

Control over energy demands in buildings may be exercised in three ways.

Meeting needs. Energy is used in a building for lighting and equipment, cooking, hot water, heating and cooling. Many of these tasks are indispensable, necessary both for survival and for acceptable levels of comfort. But the same need can be met by a variety of solutions, each with a different energy demand. Therefore energy demands can be reduced or eliminated altogether by defining how needs are met. For example, it may be possible to design to avoid air conditioning or to reduce the need for mechanized access such as elevators and escalators.

Meeting needs efficiently. Energy supply should meet user needs efficiently. For example, the building heating and cooling system must be designed, maintained and operated to keep temperatures at desired levels with minimum energy inputs and wastage. Waste energy can be used to meet the energy demands of other functions, such as heat recovery from exhaust air.

User awareness. User demands for energy must match their own needs. Users must therefore have the awareness to demand energy-consuming services only when required, and be given the means by which to exercise this control. For example, lighting should be on only when daylight levels are unacceptable, turned off when no longer required, and adjusted to suit individual user preferences, room function and room occupancy patterns.

meeting needs

control

user
awareness

meeting needs
efficiently

Above
Controlling demands through meeting needs, meeting needs efficiently and user awareness.

See also section three Human Factors "Control", section six Architecture "Operation", and section nine Monitoring Systems.

controlling technology

Technology is the practical application of science to realize specific purposes. Technology in buildings appears in many different types and forms, such as special materials providing insulation, fuel-burning systems generating heat, and cable networks distributing electrical power and data. Buildings often consume more energy than necessary because their technological systems are inefficient, over-specified and without effective means of control. This is especially applicable to building services such as lighting, heating and air conditioning systems.

Technological systems vary in complexity. The most complex systems (often called "advanced technology" or "high tech") are sometimes thought - erroneously - to be superior simply because they are recent innovations. In practice, less complex systems ("low tech") can often perform equally well or better. Existing technology should be efficient before more technology is introduced. For example, it is better to have an efficient boiler than an inefficient one used in conjunction with a heat recovery system.

No single level of technology is either universally or absolutely appropriate. Selecting, installing and using technology is a process which balances availability and access, performance, relative needs and costs. Energy-efficient design examines all forms of technology used in buildings to arrive at the most feasible solution to the problem in hand.

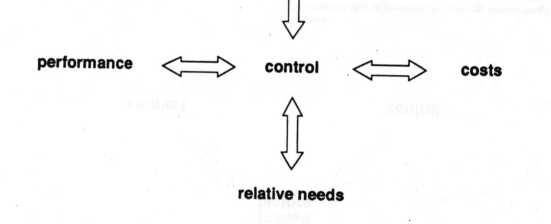

availability
and access

performance   control   costs

relative needs

Above
Controlling technology through availability and access, performance, relative needs and costs.

See also section three Human Factors "Control", section four Environment "Environmental concerns", section five Economics "Capital and finance" and "Payback and running costs", section six Architecture "Size, type and function" and "Operation", section seven Building Elements "Cooling systems", "Heating sources", "Heating systems", "Lighting" and "Ventilation", section eight Solar Applications "Cooling" and "Heating", and section nine Monitoring Systems.

human factors

This section, *Human Factors*, explains three aspects of the human condition as they occur in buildings and their technological systems: comfort (the perceived acceptability of the physical environment), control (controlling the physical environment to meet comfort, personal needs and energy efficiency), and health and safety (the physical and psychological well-being of people, and their freedom from the risk of physical harm).

The section identifies how air, temperature and light conditions must be designed and maintained for each of these three factors. Reference should also be made to *section six Architecture, section seven Building Elements* and *section eight Solar Applications*.

In many cases, statutory legislation and best practice guides will provide specific performance standards for individual countries, building types and tasks. Detailed advice and prescriptions should be obtained from these sources.

**health
and safety**

**human
factors**

comfort   **control**

Above
Human factors to be considered in the energy-efficient design process.

comfort

Comfort is a measure of human acceptability of the physical environment. In buildings, comfort is the perceived acceptability of air, temperature, light noise and so forth.

Because comfort is a subjective experience which varies between people, across countries and from season to season, it is measured by minimum and maximum (not absolute) standards intended to ensure the comfort of as many people for as much time as possible.

Designing buildings for comfort means considering statutory legislation and best practice guides for a number of factors: for example, air temperature must be at an acceptable level to minimize energy demands for heating or cooling, and should not vary greatly over a room; air and surface temperatures should not differ greatly; air movement should be appropriate for cooling and heating conditions; high humidity and high temperatures should be avoided in combination; and light should be controlled to prevent glare and excessive heat gains and so reduce energy demands for cooling.

Predictions must be made for the overall building design and for individual building elements - particularly cooling and heating systems, lighting, ventilation and windows. Because comfort varies according to activity, clothing and personal needs, (in general, people find the lowest temperature within their own comfort range to be the most acceptable), providing individual control is also advisable. This is especially important for buildings using solar systems and which can be prone to large temperature fluctuations.

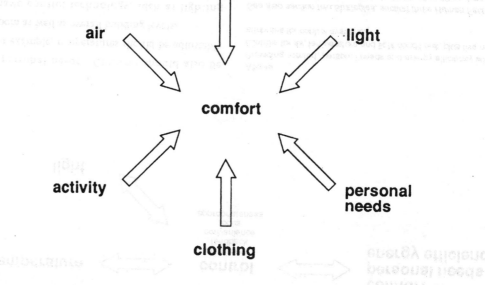

Above
Six contributing factors to comfort in buildings.

See also section two Strategies, section three Human Factors "Control" and "Health", and section six Architecture "Operation" and "Laws and regulations".

15

control

Energy-efficient design gives building users the ability to control the building and its technology. In order to reconcile energy efficiency with comfort and personal needs, users must some control over heat gains and losses, energy demands, and the technological systems incorporated into the building.

Controls allow users to adjust building performance levels for air, temperature and light conditions: for example, cooling and heating systems should have thermostats and timing devices to set temperatures and operating hours; windows and lighting systems should be adjustable for appropriate light conditions without excessive glare or heat gains; and windows and ventilation systems should be adjustable. Controls are especially important for buildings using solar systems and which are prone to large temperature fluctuations, or, for example, excessive heat gains will create energy demands for cooling.

Control systems should provide accurate feedback to users so they know what performance level is set, and should be conveniently adjustable to energy efficiency

goals and personal needs. Controls should also be focused: for example, temperatures should be adjustable for single room as well as overall building levels.

Automatic control technology, such as lighting controls responding to building occupancy, must also meet user needs and be adjustable as required. Electronic programmable control systems can offer this combination of independent operation with user intervention as required, but need careful specification to suit building type and operation pattern.

air

temperature ⟺ control ⟺ comfort
adjustment personal needs
feedback energy efficiency
convenience
focus
appropriateness

light

Above
Providing comfort, personal needs and energy efficiency with controls for air, temperature and light conditions, plus five necessary attributes for control systems.

See also section two Strategies, section three Human Factors "Comfort" and "Health", section six Architecture "Size, type and function", "Operation" and "Laws and regulations", and section nine Monitoring Systems.

health
and safety

Health is the physical and psychological well-being of people. It is a positive state and not simply the absence of illness.

In buildings, healthy conditions are primarily determined by the characteristics of air, temperature, light. Although sometimes important for health, other factors such as noise, ambience and colour are normally less important for energy-efficient design. Designing for health in conjunction with energy-efficiency therefore means considering the following.

Air. Ventilation must provide adequate supplies of fresh air, while maintaining comfortable humidity levels and minimizing pollutants. Certain types of building, room and task require mechanical ventilation or air conditioning, with attendant energy demands.

Temperature. Heating and cooling systems must maintain temperature levels acceptable for comfort and individual needs. Temperature levels are usually set between limits defined by statutory legislation or best practice guides.

Light. In general, people prefer natural to artificial

light, and prefer a view to the outside of the building or into another a room where possible.

Safety is freedom from the risk of physical harm. In buildings, safety must be ensured for the use of the building and its technological systems, particularly those with the potential of causing harm from mechanical operation, electrocution, noise, heat or disease. In general, designing for safety in conjunction with energy-efficiency means considering all relevant statutory and best practice guides.

air
air flow
humidity
pollutants

health

temperature
individual levels
comfort levels
statutory levels
best practice levels

light
natural light
artificial light
view

Above
Three determinants of healthy conditions in buildings.

See also section two Strategies, section three Human Factors "Comfort" and "Control", section four Environment "Environmental concerns", and section six Architecture "Operation" and "Laws and regulations".

environment

The environment is the natural physical world on earth. This section, *Environment*, explains two aspects of the physical world as they affect and are affected by buildings.

Prevailing weather conditions, usually classified within climatic zones or types, are one of the most important considerations of the energy-efficient design process: any successful design must take account of expected levels and variations of solar radiation, air temperature, cloud, humidity, wind and precipitation. The local effects of pollution should also be considered.

Environmental factors have always been a consideration of good design: landscaping and site planning play an important part in many architectural and building traditions. The emergence of environmental concerns on a global scale means that even greater attention is paid to such matters. Because much of this has to do with the consumption of combustible fuels, the implications for energy-efficient design are obvious and immense. Because climate and environmental concerns differ

greatly across countries, between buildings and over time, it is the responsibility of designers to obtain appropriately detailed and reliable data and other forms of information. For example, designers must discover the likely solar radiation and temperature levels and balance these against the environmental effects of a central heating and cooling system before deciding whether a air conditioning plant is really necessary.

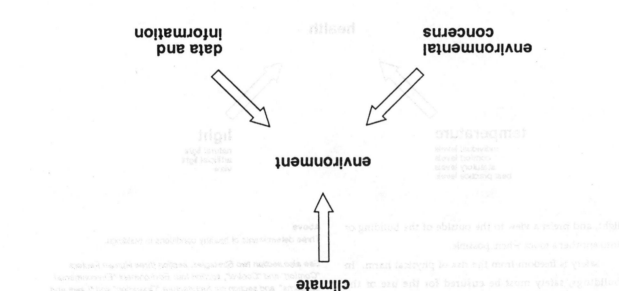

Above
Three factors to be considered when designing for the environment from the perspective of energy-efficiency.

climate

Climate refers to the prevalent and predictable meteorological conditions of a geographical area. It can be considered on three scales.

At the largest scale, a macro climate is determined by global patterns of latitude and geography. For example, high-latitude climates have pronounced variations in temperature and solar radiation between seasons; lower latitudes tend to have less seasonal variations and more temperate and tropical climates; and maritime climates tend to be milder and more humid than those with continental locations. Within a macro climate are the varying meso climates, covering smaller geographic areas, determined by regional land and water topography. At the smallest scale, micro climate is determined by local factors (such as building layout, soil and vegetation) which modify the regional or meso climate.

Designing for climate from the perspective of energy-efficiency means considering the following factors.

Solar radiation. Solar radiation levels and patterns are determined by latitude, cloud cover and humidity. Higher levels of solar radiation mean lower energy

demands for heating and artificial lighting, and higher energy demands for cooling. All kinds of climate can support some form of exploitation of solar radiation, although purpose and use will vary: for example, solar radiation can be used for space and water heating in temperate climates, and for water heating only in hot climates. In cool climates, the building should be orientated to maximize solar radiation, and surface materials can be chosen which help reflect solar radiation

continued

cold

characteristics

low average temperatures in all seasons

lower winter solar radiation

high winter relative humidity

design implications

maximize insulation

exploit orientation

use of solar applications requires auxiliary heating systems in the winter

temperate

characteristics

high average summer and low average winter temperatures

variable temperatures in all seasons

design implications

provide insulation for winter conditions

exploit building elements and solar applications to reduce winter demands for heating

exploit building elements and solar applications to reduce summer demands for cooling

exploit orientation and landscaping to balance the effects of winter and summer

hot and dry

characteristics

high average temperatures in all seasons

high solar radiation in all seasons

low relative humidity in all seasons

high diurnal temperature range

strong winds

design implications

minimize day-time heat gains

use insulation to stabilize day-time temperatures

exploit orientation for cooling by natural ventilation

provide cooling by evaporation

provide cooling by air conditioning in extreme conditions

consider thermal capacity

hot and wet

characteristics

high average temperatures in all seasons

high relative humidity in all seasons

high precipitation in all seasons

design implications

exploit orientation for cooling by natural ventilation

limited opportunities for cooling by evaporation

consider lightweight construction with shading

Above
A simple classification of climate by four types, showing basic meteorological characteristics and the major implications for design in each case.

See also section four Environment "Environmental concerns", section six Architecture "Location", "Site", "Orientation", "Plan", and "Form", section seven Building Elements and section eight Solar Applications.

into buildings. In hotter climates, solar radiation should be minimized by adjusting orientation, exploiting available over-shadowing, minimizing window area and using lighter coloured finishes.

Air temperature. Air temperature varies both annually and diurnally. Local changes are caused by terrain, particularly hills and lakes. In general, urban areas tend to have higher temperatures than rural areas because of the higher absorption of solar radiation by surface materials. Higher average external (ambient) air temperature levels mean lower energy demands for heating and higher energy demands for cooling, and vice-versa. The diurnal temperature range is important for the provision of energy storage, and the annual temperature range is important for the provision and specification of heating and cooling systems.

Cloud cover. Clouds are formed as air is cooled and moisture collects around particles in the air. This blocks the passage of sunlight and diffuses solar radiation. Consistent or frequent cloud cover therefore reduces the effectiveness of solar collection systems, particularly those active systems which depend on direct solar radiation.

Humidity. Humidity is water vapour in the air. Whether a climate has a high or low of humidity level affects solar radiation levels: a humid climate means lower day-time solar gains and night-time radiation losses, and an arid climate means higher day-time solar gains and night-time radiation losses.

Wind. The natural movement of air increases energy demands for heating and may reduce energy demands for cooling. Buildings in temperate and cold climates should be designed to reduce surrounding wind speeds and thus reduce heat losses through infiltration, conduction and convection. Exploiting site conditions and landscaping can greatly reduce wind speeds. Building orientation in hot and wet climates should be designed to exploit local wind patterns and thereby provide cooling.

Precipitation. Precipitation occurs as rain, dew, hail, snow or sleet. Precipitation patterns determine roof and wall designs. In general, high or year-round precipitation requires steeply pitched roofs together with walls designed to facilitate the drying-out process, and low or seasonal precipitation allows the use of flat or low pitched roofs. Precipitation on the outer surfaces of a building increases heat losses by conduction and evaporation and has a detrimental effect on the general condition and performance of a building; higher insulation and construction standards may therefore be required.

environmental concerns

Environmental concerns deal with the impact of human actions upon the surrounding environment. This includes changes to meteorological conditions, depletion of natural resources such as forests and fuel supplies, and degradation of natural habitats for indigenous peoples and wildlife. In particular, global warming is linked to the presence in the atmosphere of carbon gases from combustion, and the depletion of the ozone layer from the emission of chlorofluorocarbons (CFCs). More solar radiation passes through the atmosphere and more heat emitted by the earth is trapped, causing a gradual but significant rise in global temperatures. Environmental concerns may be included in the energy-efficient design process in four ways.

Design. In general, designing architecture to maximize the benefits of solar radiation and natural ventilation reduces energy demands for heating and cooling, and including landscaping in designs means more carbon dioxide is absorbed. The design process should include continual monitoring and feedback to ensure that the environment is properly considered

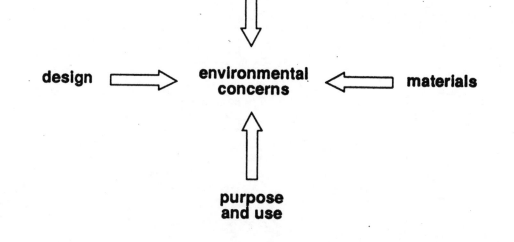

throughout the building's life.

Materials. The effects of CFCs may be reduced by avoiding the insulants, aerosols, cleaning fluids and refrigeration equipment that contain CFCs. Pollutants, particularly oxides of nitrogen and sulphur, carbon monoxide and fuel gases, can be restricted through appropriate specification of heating and cooling systems. Non-toxic materials that are easily maintained, replaced, disposed of, and that do not deplete natural resources

continued

Above
Four factors to be considered when designing for environmental concerns from the perspective of energy-efficiency.

See also section one Energy and Efficiency "What is energy?", section three Human Factors "Health and safety", section four Environment "Climate", section five Economics, section six Architecture "Site", "Construction and materials", "Operation" and "Laws and regulations", and section seven Building Elements "Cooling systems", "Heating sources", "Heating systems", "Landscaping" and "Ventilation".

should be used. For example, these include recycled paper products and wood from sustainable sources.

Technology. Buildings should be aim to be natural, simple and resourceful rather than highly technological, over-designed and overly complex. For example, natural light and ventilation are preferable to artificial lighting and air conditioning, and buildings should be adaptable for a long useful life without undue built-in obsolescence. Technology is best used in support of passive design, or where its impact is clearly beneficial or harmless to the environment.

Purpose and use. New buildings should only be constructed where there is sufficient purpose. Re-using or adding a use to an existing building usually requires less energy and materials and produces less pollution than constructing a new building. In terms of using buildings, environmentally-sound architecture is a matter of personal responsibility, with individuals acknowledging that small-scale decisions and actions can collectively produce large-scale effects. For both designers and users, disseminating knowledge and having an awareness of the world outside of the building are therefore of central importance.

economics

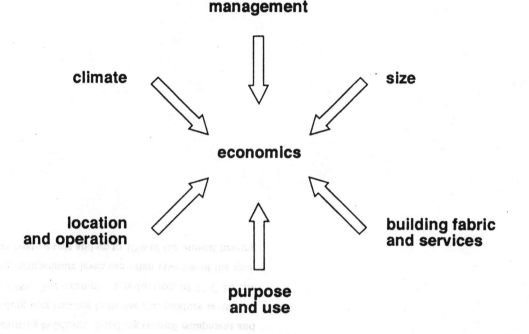

Through energy-efficient design, economic savings are made in the energy consumed during a building's life. Because running costs (costs-in-use) are lower than for a conventional building, after a period of time (payback period) any extra design and construction costs (capital costs or first costs) are recovered. Savings are then made in running costs for the rest of the building's life. A short payback period therefore means greater cost-effectiveness. Six factors determine the economics of energy-efficient design.

Purpose and use. New buildings should only be constructed where there is sufficient purpose. Re-using or adding a use to an existing building requires a small amount of the energy and materials needed for a new building. Use is also important: a long-life building should be adaptable, and a short-life building should have materials chosen for the potential for re-use.

Location and operation. More energy can be consumed as a result of building location and operation, such as moving goods and people, than is consumed in the building itself. Because these costs are often incurred by people renting or using the building, and not by the landlords or those who built it, they are rarely taken into account. Nevertheless, the energy and environmental costs to society as a whole should be considered.

Climate. Free solar energy is relatively easy to obtain, even at northern latitudes. The problem for designers is to make sensible use of it inside the building. A climatically-interactive building exploits solar radiation, natural light and natural ventilation. The main

continued

Above
Six factors which determine the economics of energy-efficient design.

See also section four Environment and section six Architecture.

economic effect is to reduce both the length and severity of the winter heating season and summer cooling season and so reduce energy demands for heating and cooling.

Size. A building should not waste energy by being over-sized or inefficient in the use of space. Conversely, buildings with over-dense occupation will suffer reduced occupant performance, productivity and building flexibility. Running and capital costs should be assessed per unit occupant and per unit square metre of usable floorspace to allow cost comparisons to be made between buildings.

Building fabric and services. Energy savings may be made by controlling the design of the building fabric, materials and elements. For example, adjusting orientation and south-facing glazing can easily lead to 10% reductions of energy demands for heating in a dwelling. The energy costs of materials should be considered, especially for the potential for re-cycling and re-use. Many materials offer a use advantage as well as an energy saving. Of the various building elements, relatively small investments in insulation can give especially good improvements in environmental conditions coupled with good economic returns, depending on a cost-benefit threshold between building design, its pattern of use, and energy prices.

Running costs for services can be reduced by using efficient systems and good control equipment which is conveniently located, easy to understand and not overly complex can repay its cost within a few months.

Centralized (also known as district or remote source) heating schemes are often inefficient in operation due to distribution losses, wasteful behaviour by users, lack of metering, and uncontrolled pricing. They may nevertheless be the correct choice if they can make use of low-cost energy, particularly "waste" heat from electricity generation.

Management. Management can be effective in new and existing buildings. Installing control equipment and providing user training in its use can produce savings of over 25%. For example, a reduction of 1° C in the average temperature levels can often save 5% of the mid-winter heating cost and up to 10% of the annual heating cost.

capital and finance

Capital is needed to design and build a building. Energy-efficient buildings often - although not always - cost more to design and build. The size of this overcost differs between countries, buildings and designs. In general, overcosts average 10% of design and construction costs within a range of 0-25%, and are lower for designs using conventional construction, materials and technology.

Finance of overcosts may come from a number of sources. *Building clients* may bear overcosts where they will benefit from energy savings or where overcosts can be passed on to building owners or tenants. *Designers* may bear overcosts where the building is for their own use. *Governments* at all levels may finance overcosts, particularly for programmes such as the use of insulation. *Research institutions* may finance designs which focus on the same research area as their own. *Professional institutions* may finance designs which promote the profession. *Private sector* manufacturers and commercial interests may finance overcosts where designs demonstrate the value of their products or services. *Private patrons* may finance designs, particularly where

other sources are unhelpful. *Financial institutions* may finance total building costs, although the building must be worth the amount loaned or collateral may be required. *Charities and development agencies* may finance total building costs or energy overcosts, and can be for designs supplied by themselves.

Those who pay overcosts do not always benefit from energy-efficient design. Overcosts are therefore often passed on as increased purchased costs for building owners and increased rents for building tenants.

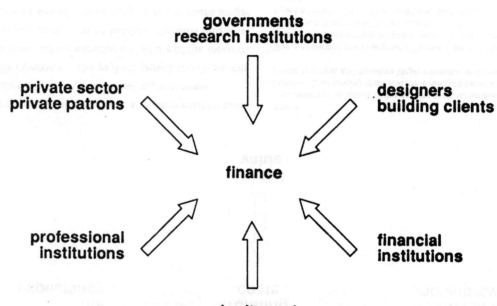

Above
Possible sources of finance for the overcosts of energy-efficient design.

See also section five Economics "Payback and running costs" and section six Architecture "Construction and materials".

payback and running costs

The payback period is the time it takes for the overcosts of energy-efficient design to be repaid as reduced running costs. For long-life measures, such as increased insulation, a payback period of 10-20 years may be considered reasonable. Shorter-life measures, particularly high-technology measures such as mechanized shading devices, need to pay for themselves more quickly. The most important running costs are as follows.

Energy. The amount and cost of energy consumed is the main determinant of running costs. Lower energy consumption and rising energy prices mean relatively lower running costs compared with conventional buildings, and therefore shorter payback periods.

Maintenance and improvement. Designs for conventional construction, materials and technology, such as exploiting orientation for solar radiation and light, require less attention. Complex technology is often less successful: for example, mechanical ventilation systems with heat recovery and movable shutters can lessen energy efficiency or fail if badly maintained. As many energy-efficient buildings are

experimental or one-off designs, allowance must be made for higher maintenance levels and improvements.

Life expectancy. The payback period should usually be less than half the expected overall life of the measure. High replacement costs of building elements and solar applications also add to running costs and mean longer payback periods.

Added value. Positive factors which cannot be easily costed should also be considered: for example, sunspaces raise internal quality, amenity and property value.

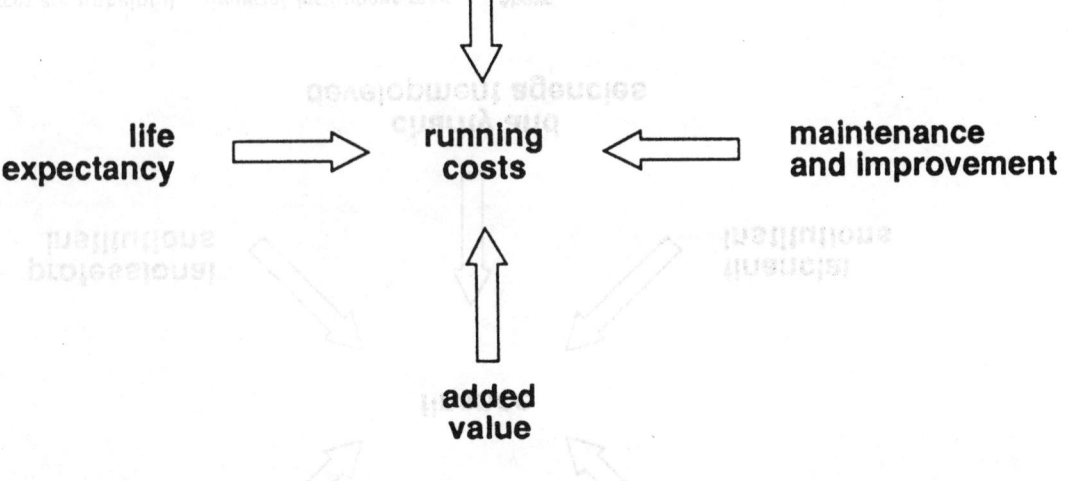

Above
Four determining factors of the running costs for an energy-efficient building. Low running costs and low overcosts mean a short payback period, and therefore larger economic and energy savings.

See also section one Energy and Efficiency, section two Strategies, section four Environment "Environmental concerns", section five Economics "Capital and finance", section six Architecture "Size, type and function" and "Operation", and section nine Monitoring Systems.

part two

defining the brief

architecture

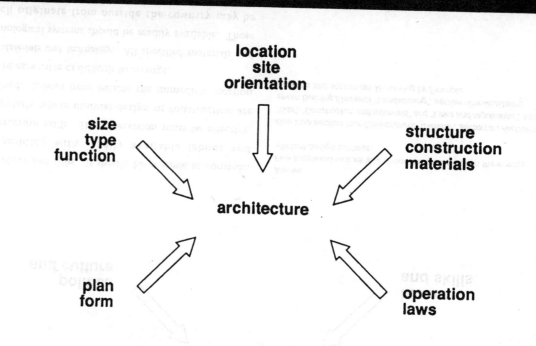

Architecture is a process - the art and science of designing buildings - and also a product - the buildings which this design process creates. This section, *Architecture*, therefore embraces the design process, the construction process, the building as it is intended to be used, and the building as it actually performs and is used in practice.

Together with the other sections in *Part Two* of this guide, *Architecture* makes up an instructive framework for design. Information is given in the form of basic principles and general recommendations to be adapted to individual buildings as required.

The focus is on the most important factors which influence design: those factors which affect decisions about the building as a whole. This includes where the building is located, positioning on site, the direction it faces, arrangement of internal rooms, building size, building type, building function, the primary structure, construction methods, material specification, the building in use, and legal obligations for the designer and building. Each of the above is explained from the perspective of energy-efficient design to allow new

designs to be explored and existing designs to be verified for energy efficiency. The design of individual components and building details may then be undertaken with a full understanding of the larger energy-efficiency framework.

More detailed information about these components and details may be found in *section seven Building Elements* and in *section eight Solar Applications*. Reference should also be made to *section one Energy and Efficiency* and to *section two Strategies*.

Above
Thirteen factors which affect design decisions about the building as a whole.

29

ARCHITECTURE

location

Location is the geographical position of a building, and determines whether a building design can be successfully implemented. Designing for location from the perspective of energy efficiency means being able to make informed decisions that take into account a number of locational factors.

Climate. Air temperature, solar radiation, humidity, wind, cloud and precipitation data for the surrounding geographic region should be obtained. This is critical for non-conventional designs and solar applications.

Politics and culture. The overall design of the building should be appropriate for the general political and cultural expectations of the area. Unusual design may make permission to build difficult to obtain.

Knowledge. The knowledge with which to design, build and operate a building should be available. For example, advanced design methods may be required to predict the likely effects of an atrium space on heat gains and ventilation patterns, and a good understanding of these effects during operation may be required to know when to provide shading and increase ventilation.

Labour and skills. It should be possible to construct the building with readily available labour and construction skills. Site supervision must be effective, particularly where unusual design or construction are involved. Labour from outside the immediate location may be expensive or difficult to arrange.

Materials and technology. All specified materials and technological systems should be readily available. Those which originate from outside the country may be expensive or difficult to procure and maintain.

labour and skills

politics and culture

knowledge

location

climate

materials and technology

Above
Five locational factors that should be considered in the energy-efficient design process.

See also section four Environment 'Climate', section six Architecture 'Site', 'Construction and materials' and 'Laws and regulations', section seven Building Elements 'Landscaping', section nine Monitoring Systems and section ten Monitoring by Example.

site

Site refers to the natural terrain immediately beneath and around a building. Designing for site conditions from the perspective of energy efficiency means considering the following factors.

Solar radiation and natural light. Although expected solar radiation and natural light levels for a given location may be modified by slope and over-shadowing from surrounding buildings and landscape, this is unlikely to prevent exploitation of solar radiation as long as the climate in general is appropriate.

Greenery. Trees and other vegetation help reduce solar radiation, wind, pollution, noise and air temperatures, and increase humidity. In general, small 5-10m trees south of the building provide shade without drastically reducing solar radiation. Shelter from prevailing winds may be provided by intermittent trees and shrubbery. Greenery may also modify the thermal characteristics of the soil type underneath.

Wind. Wind increases heat losses by infiltration and by convection and conduction of heat away from the building surface, and disperses solar gains held in the external air. Where heating loads are significant, exposed sites should be avoided and other buildings and greenery should be exploited for shelter. Where cooling loads are significant, winds across the site may be exploited to provide ventilation, especially at night when wind speeds tend to rise.

Wind may determine roof design: windward pitched roofs below 30° experience negative pressures and those above 30° experience positive pressures from the

continued

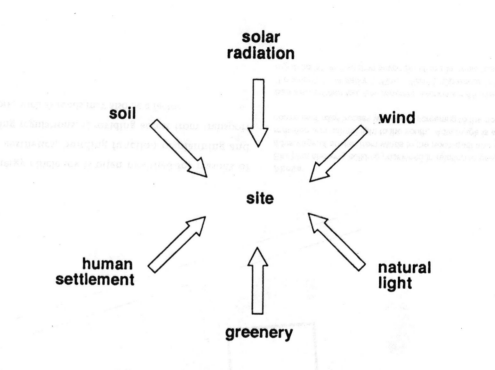

Above
Six factors to be considered when designing for site conditions from the perspective of energy-efficient design.

surrounding air.

Winds may be used to power active devices such as windmills which generate electricity and pump water for buildings, although this may conflict with the need to provide shelter. Such devices may therefore need to placed away from the building on the site.

Soil. Soil characteristics affect heat losses from buildings and the potential for thermal storage in the ground. Dry sandy soil types have the greatest diurnal temperature range and the quickest thermal response. Wet compact soils demonstrate more stable thermal characteristics, and often require insulation to be added to the building ground floor.

In exposed conditions, soil may be used in banks around walls as an additional insulation layer and thus reduce heat losses from conduction, convection and infiltration. Soil may provide the source for fuels in the form of combustible wood, and in some locations, soils such as peat may itself be used as a fuel. In suitable conditions, heat in the soil may be used to pre-heat or pre-cool ventilation air, or may be used as a source/sink for a heat pump.

Human settlement. Other buildings, especially in densely developed urban sites, create humidity variations from altered natural drainage patterns, lower solar radiation from airborne pollution and over-shadowing, and reduced wind speeds and slightly higher overall temperatures from building clustering.

As with orientation and plan, a site design optimized for energy efficiency is often modified for reasons of view, aesthetics, building function or planning and building regulations. Providing access from transport networks such as roads may also be a factor.

Above
Site plan showing building positioned in relation to greenery to take advantage of shelter from winds to the north-east and of solar radiation and natural light to the south. Advantage is also taken of convenient road access with some screening to the south.

See also section four Environment, section six Architecture "Location", "Orientation", "Plan", "Form", "Operation" and "Laws and regulations", and section seven Building Elements "Landscaping"

orientation

Orientation is the direction of the building plan relative to north and south, and to the sun and wind patterns. In the northern hemisphere, the building's south side experiences increased heat gains from solar radiation and natural light, and the building's north side experiences lower heat gains and natural light and possibly greater heat losses from exposure to wind and precipitation. In the southern hemisphere these relations are reversed.

Designing for orientation from the perspective of energy efficiency means exploiting solar radiation, natural light and ventilation, and avoiding winds, precipitation and, in cold climates, over-shadowing.

For the building south side, large window areas maximize the potential of solar radiation and natural lighting, but may require multiple glazing and shading devices to control heat losses, heat gains and glare according to season and climate. In general, slight increases in heat loss are acceptable if energy demands for artificial lighting can be reduced.

For the building north side, high wall insulation values and smaller multiple-glazed windows may be required. Fragile building elements such as active solar collector panels should be in sheltered positions. Where cooling is required, building orientation should exploit wind patterns and increase natural ventilation.

As for site and plan, orientation optimized for energy efficiency is often modified for view, aesthetics, site conditions, building function or planning and building regulations. Some deviation from the south, possibly up to 30°, may be made without significantly increasing heating requirements.

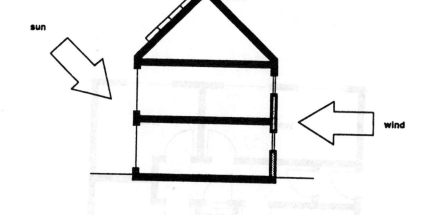

sun

wind

Above
Section showing design optimized for orientation. Large single-glazed windows on the south side increase natural light and heat gains from solar radiation. Small double-glazed windows and insulated walls on the north side reduce heat losses. Active solar collector panels on the south face of the inclined roof are turned towards the sun and are sheltered from wind.

See also section four Environment, section six Architecture "Site", "Plan", "Form" and "Laws and regulations", section eight Building Elements "Insulation", "Landscaping", "Sunspaces" and "Windows", and section nine Solar Applications "Heating".

plan

A building plan is the arrangement of internal spaces. Designing a plan from the perspective of energy efficiency means considering the influence of room position and inter-connection on heat losses and heat gains.

Where heat losses are significant, internal spaces can act as a buffer: for example, non-heated rooms on the north side protect heated rooms to the south; a sealed staircase reduces heat losses from heated ground floor to unheated upper floor rooms; and a hallway protects heated spaces from the outside.

Where temperatures are high, such as temperate and hot-humid climates, arranging rooms, doors and windows for cross-ventilation reduces energy demands for cooling.

For windows on opposite walls, more ventilation occurs when the wind is less than 90° to the windows; for windows on adjacent walls, more ventilation occurs when the wind is at 90°. An inlet lower than the outlet means more even ventilation. Smaller outlets and larger rooms mean slower ventilation.

Rooms needing heat and light should be on the south side to exploit solar radiation and should not be over-shadowed by non-heated rooms. Rooms used in the morning should be on the east side to benefit from the sun. Services should be in concentrated areas to reduce costs and increase efficiency.

As for site and orientation, a plan optimized for energy efficiency is often modified for view, aesthetics, site conditions, building function or planning and building regulations.

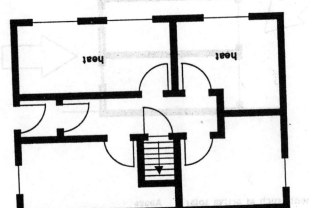

Above
Ground floor plan showing internal spaces arranged from the perspective of energy efficiency. Heated rooms on the south side are protected by non-heated rooms on the north side. An internal porch reduces heat losses through the main entrance, and a door to the staircase reduces heat losses to upper floor rooms.

See also section four Architecture "Site", "Orientation", "Form" and "Laws and regulations", and section eight Building Elements "Doors", "Sunspaces", "Ventilation" and "Windows".

form

Form is the shape and appearance of a building. Designing form from the perspective of energy efficiency means considering wall length, plan area, perimeter complexity, building height, window pattern and aesthetic effect.

In general, compact shapes with increased floor area and volume in relation to surface area are the most energy efficient due to reduced heat losses. Wide and shallow forms can increase heat gains from solar radiation through a long south side, but increase the potential for heat losses through the north side at night and in cold weather. Conversely, narrow and deep forms in terraces can reduce heat losses from individual units but also reduce exploitation of solar radiation. A trapezoidal form with a long south and short north side may increase heat gains and reduce heat losses. Complex forms may cause over-shadowing, reduce heat gains from solar radiation, and increase heat losses from infiltration.

Energy efficiency of building shapes is closely related to window patterns. Windows in walls usually offer lower heat losses than roof skylights, although choice is

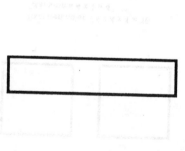

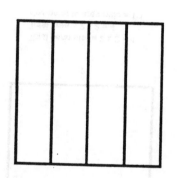

often determined by room function and building type. Compact shapes using large areas of south-facing glazing without provision for shading and ventilation can cause overheating or increase energy demands for cooling. Heat gains also may be controlled by colour: dark surfaces tend to absorb and light surfaces tend to reflect solar radiation.

The acceptability of building aesthetics and style is dependent on local building traditions and social culture. Active solar systems often require careful integration.

Above
Four kinds of building shape in plan (clockwise from top left): trapezoidal, complex, terrace, and wide and shallow.

See also section six Architecture "Orientation", "Plan" and "Laws and regulations", and section seven Building Elements "Sunspaces", "Ventilation" and "Windows".

size, type and function

Size, type and function are simple categories of building. Designing for them from the perspective of energy efficiency means understanding basic relationships between buildings, human needs and energy.

Making a building smaller reduces capital and energy costs of construction, and reduces energy demands for heating and lighting. However, this depends on a small building meeting the same needs as a larger one. Small buildings can cost more to construct and consume more energy than a large building due to materials and systems duplication, and due to higher heat losses from increased surface area in relation to floor area and volume. Close grouping of buildings reduces heat losses but can also reduce the exploitation solar radiation, natural light and natural ventilation offered by a larger perimeter.

Energy-efficient design must be appropriate for building size and type. Heating systems in particular should be specified to meet the size and pattern of heating loads. In general, exploiting solar radiation and natural light is more appropriate for domestic and small buildings and less appropriate for non-domestic larger

buildings, particularly those with deep plans and high service requirements. Some fuel and building combinations may be prohibited: for example, gas supplies in tall structures.

Building function is therefore a primary factor in energy-efficient design: for example, many offices use complex cooling, heating and lighting systems to produce uniform conditions across rooms; and schools needing quick thermal response and high light levels may be lightweight structures with large windows.

Above
Four small buildings compared with a single large building with the same total floor area. Because the four small buildings have a larger perimeter, the potential for heat losses is increased. However, the opportunity to exploit heat gains from solar radiation, natural light and natural ventilation may also be increased.

See also section five Economics, section six Architecture 'Form', 'Structure', 'Construction and materials', and 'Operation'.

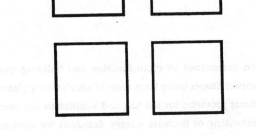

1
Total perimeter = 4 x 4 x 1 = 16
Total area = 4 x 1 = 4
Perimeter to area ratio = 4:1

2
Total perimeter = 2 x 2 = 4
Total area = 2 x 2 = 4
Perimeter to area ratio = 1:1

36

structure

Structure is the part of the building which makes it stand up. Structure weight and type affect building thermal performance. Lightweight buildings using, for example, thin masonry walls and wood internal partitions store less heat and respond more quickly to solar radiation and temperature levels. Heavyweight buildings using, for example, thick masonry walls and concrete floors can store more heat and respond more slowly. With correct specification, both may offer the same insulation performance. Designing for structure from the perspective of energy efficiency means considering structure weight and type, occupancy and solar radiation.

For climate, a heavy structure is useful where heat must be stored (as for passive solar systems) or night cooling exploited (as for hot arid climates). Occupancy may also be a factor: for example, an office building unused at night must respond quickly to heating and cooling demands, which may suggest a light structure; alternatively, a hospital requires near constant internal conditions, which may suggest a heavy structure.

Solar radiation is also important: where the building

is intended to stabilize or minimize the effects of solar radiation, a heavy structure absorbs direct heat gains releases them slowly into the building interior; where the building is intended to to respond quickly or maximize the effects of solar radiation, a light structure with thin walls and large areas of glazing increases direct heat gains and transmits them quickly into the building interior.

In general, light structures are useful where internal or external temperature fluctuations are insufficient to make weight structures advantageous.

sun

low storage
quick response
high solar gains
high temperature range

sun

high storage
slow response
low solar gains
low temperature range

Above left
Section showing the main thermal effects of a lightweight structure with large areas of glazing.

Above right
Section showing the main thermal effects of a heavyweight structure with small areas of glazing.

See also section three Human Factors "Control", section six Architecture "Size, type and function", "Construction and materials" and "Operation", section seven Building Elements "Doors", "Floors", "Roofs", "Sunspaces", "Walls" and "Windows", and section eight Solar Applications "Cooling" and "Heating".

construction and materials

Construction is the process by which buildings are built. Materials are the substances and components with which a building is constructed. Designing for construction and materials from the perspective of energy efficiency means considering their availability, implementation and total energy costs.

To implement energy-efficient design, all necessary knowledge, labour, construction skills, materials and technology must be available. For example, a specialized integrated wall cladding, insulation and window system may be more difficult to design for, obtain and install than conventional building elements designed to exploit solar radiation, natural lighting and natural ventilation.

Non-conventional construction methods, materials, building elements and solar applications must be appropriate for local resources. Advantages can include easy retro-fit application to existing buildings and system adjustability after installation, but all non-conventional techniques must be carefully monitored for energy efficiency and possible dangers to health and safety.

For parts of a building not directly related to energy,

high quality specification and construction may reduce heat losses, particularly those related to infiltration and insulation. These two factors also have a long-term effect upon overall building condition and its ability to operate as an energy-efficient system.

Where possible, the energy used to produce materials and construct a building should also be considered. For example, a laminated wood beam may require only 20% of the energy for the extraction, processing and installation of a steel beam of equivalent strength.

total energy costs

construction and materials

availability

implementation

Above
Three factors to be considered when designing for construction and materials from the perspective of energy-efficient design.

See also section three Human Factors "Health and safety", section four Environment "Environmental concerns", section five Economics "Capital and finance", section six Architecture "Size, type and function", "Structure" and "Laws and regulations", section seven Building Elements, section eight Solar Applications and section nine Monitoring Systems.

operation

Operation is the way a building performs and is used in practice. Designing for operation from the perspective of energy efficiency means understanding that energy efficiency is ultimately determined not by building design but by the amount of energy consumed.

The way buildings are used can differ from how the designer intended. Buildings must therefore be able to create a range of environmental conditions in a manner that is energy-efficient, convenient and acceptable for building users. Complex or fragile technology, such as air conditioning systems or active solar collectors, require careful integration with the rest of the building and may be difficult to operate at required performance levels.

Designers and users must be able to understand building use in practice. Feedback should be provided for things like the amount of energy consumed for the building as a whole, for particular fuels, and for specific functions such as heating and lighting. Feedback should be appropriate to building type and size, and can vary from a simple meter to inform occupants of the electricity consumption of their home to a complete

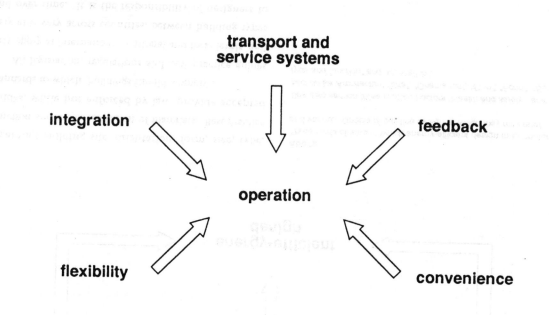

automated energy audit for an office building.

Where possible, the energy consumed outside of the building should also be considered. For example, using public transport may be encouraged by making it accessible and pleasant to use, and by giving private car parking low priority. Locating buildings closer together while preserving an acceptable micro climate reduces energy demands for transport and increases system and maintenance efficiency of electricity, gas, water, sewage and telecommunications services.

Above
Five factors to be considered when designing for operation from the perspective of energy efficiency.

See also section three Human Factors "Control", section four Environment "Environmental concerns", section five Economics "Payback and running costs", section six Architecture "Size, type and function", "Structure" and "Laws and regulations" and section nine Monitoring Systems.

laws and regulations

Statutory legislation and building regulations provide legal obligations concerning the way buildings are designed, constructed, operated and maintained. Designing for legislation and regulations from the perspective of energy efficiency means ensuring that the building and its building elements conform to all energy-related standards and that they do not prevent compliance with all other legal obligations.

Energy-related standards may include lighting levels, temperature levels, humidity levels, ventilation rates, fuel type and overall insulation performance. It is likely that these standards will become wider-ranging, more detailed and more demanding over time.

Other standards relate to health and safety, and may include structural stability, fire resistance, water penetration, electrical and gas systems, toxic substances, pollutants, air quality and ergonomics. All aspects of building design must conform to these standards. Energy-efficient design must not interfere with this process.

Planning regulations may also restrict design choices regarding building site, orientation, form, size, type, function and the specification of materials. Best practice guides, while not enforced by law, provide accepted standards to which buildings should comply.

All legislation, regulations and best practice guides may apply at international, national and local levels, and may also vary across countries, between building types and over time. It is the responsibility of designers to ascertain their own legal obligations and ensure that these obligations are fulfilled.

energy-efficient design

planning regulations
site
orientation
form
size
type
function
materials

health and safety standards
structure
fire
water
electricity
gas
toxics
air
ergonomics

energy-related standards
lighting
temperature
humidity
ventilation
fuel
insulation

Above
Three kinds of laws to which energy-efficient design must conform, and various aspects of building design to which they may relate.

See also section three Human Factors 'Health and safety', and section six Architecture 'Site', 'Orientation', 'Plan', 'Form', 'Size, type and function', and 'Operation'.

building elements

The elements which make up a building satisfy the requirements of structure, external protection, building function, user comfort, cost, security and laws and regulations as well as the requirements of energy efficiency. These are the minimum performance criteria to be met by each element individually and as part of the whole building. This section, *Building Elements*, provides a definition of each element, the performance criteria which it must meet, and its significance from the perspective of energy-efficient design.

The elements presented here are the conventional ones to be found in most buildings: doors, walls, windows and so forth. They therefore constitute a given framework within which the control of gains, losses, demands and technology may be exercised, and each building is uniquely made up from different variations and combinations of these elements. *Building Elements* provides an introduction and a reference source for this subject. Those elements designed largely with solar heating and cooling considerations in mind are dealt with in *section eight Solar Applications*.

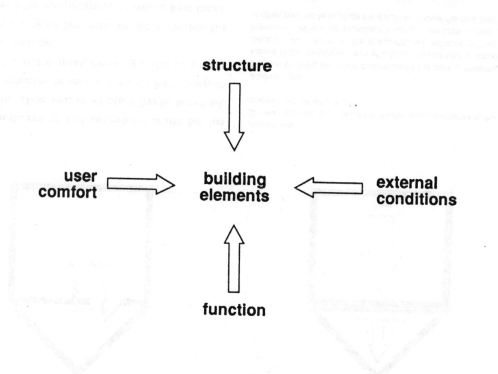

Above
Four performance criteria which building elements must meet.

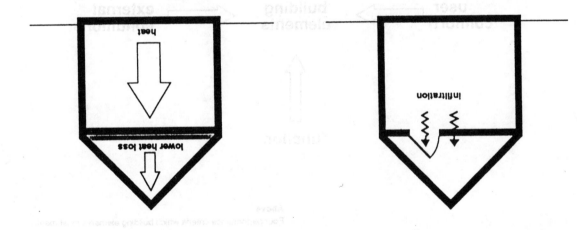

ceilings

Ceilings define the upper surfaces of rooms. They provide acoustic and visual privacy between floors, or act as an additional barrier to external elements by separating the roof space from used space below. Lighting and other services are usually located in the ceiling, especially when a suspended construction provides a ceiling void.

Ceilings can also be used for the transmission into the building of heat collected by roof ponds located in the roof space. In such cases, metal or concrete decking is often used to encourage heat conduction and radiation into the rooms below.

Ceiling insulation keeps warm or cool air within a building as required. Insulating performance depends on the thermal properties of the ceiling materials (such as timber and plaster), the insulation type used (such as mineral wools, cellulous fibres and polystyrene), and on quality of construction. Insulation may be improvised, such as from polystyrene previously used for packing.

Heat losses may occur through cracks and joints in the ceiling and around attic hatches, or through apertures in the ceiling, such as open loft or attic hatches or stairwells. These heat losses can either be beneficial, creating a reduction in energy demands for cooling, or they can be detrimental, creating a rise in energy demands for heating.

Services carrying hot water or hot air within the ceiling should be also insulated to reduce heat losses. Where desirable, incidental heat gains from lighting should be directed into the room below, rather than absorbed by the ceiling.

Above left
Section showing heat loss by infiltration through cracks in the ceiling and around the attic hatch.

Above right
Section showing insulation placed above the ceiling, reducing heat losses to the roofspace. Any attic hatch should also be insulated. Because warm air rises, this is a particularly efficient method of insulation. Where the roofspace is used or contains a water tank, it is often preferable to place the insulation below the roof itself.

See also section seven Building Elements 'Floors', 'Insulation' and 'Roofs', and section eight Solar Applications 'Cooling' and 'Heating'.

cooling systems

Cooling systems in buildings cool internal spaces. Specification depends on climate, external air quality, building form, size and function, comfort levels and capital. Three arrangements are possible.

Natural ventilation. Natural air movement, caused by differences in pressure or temperature, cools by replacing warm indoor air with cool outdoor air and by improving convection and conduction from the body. It consumes no energy, and has few if any running costs, but depends on weather conditions, is difficult to control, and cannot reduce air temperatures below ambient levels.

Isolated units. Cooling units for specific rooms are often mechanical devices, typically fans, which increase air supply, movement and extraction. Alternatively, air conditioning units with an internal refrigeration device can reduce air temperatures below ambient levels.

Central source systems. Central air conditioning systems powered by electricity are the most common way of cooling air below ambient temperatures. Air is cooled by a central refrigeration plant and distributed around the building by a system of fans and ducts. In all-air

systems cooling occurs by controlling supply air condition, and in air-and-water systems supply air is treated in a central plant and cooled at the distribution point by induction unit or by chilled water in a fan coil.

Waste in all energy-driven cooling systems is caused by inefficient air cooling, system losses and over-use. Careless design, operation and maintenance lead to very high energy consumption. Efficiency means maximizing natural ventilation and providing cooled air only as needed. Flexibility and user control are important.

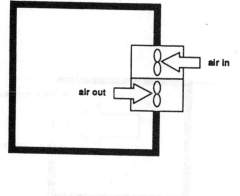

Above left
Isolated unit cooling using a simple two-way system of mechanical fan ventilation for air supply and air extraction. Fans may be placed on external walls, or connected to the outside by ducts. One-way systems which combine natural ventilation with either mechanical air supply or extraction are also widely used.

Above right
Central source cooling using an air conditioning system. Air flow rate, purity, temperature and humidity can all be controlled, and the heat pump can also be arranged for heating as required. The use of air recirculation increases energy efficiency.

See also section seven Building Elements "Heating sources" and "Ventilation", and section eight Solar Applications "Cooling".

doors

Doors are used as points of entrance and exit into a building, and between rooms inside a building. Doors are made from a variety of materials, including wood, metal, glass, plastic and combinations of these. In addition to fire, position, privacy and security considerations, door specification depends on energy-efficient design.

Heat losses also occur through infiltration between the door and the door frame are usually more important than heat losses through the door itself. Insulated doors are effective only if they are well sealed. Wood, the most commonly used door material, can warp and thereby increase infiltration and heat loss, although modern seals can reduce some of the effects of this distortion.

Doors should be located away from prevailing winds or should be screened, such as by external walls and trees, in order to reduce heat losses when open. The use of two sets of doors, self-closing mechanisms or draught-sealed revolving doors may be required if the door is in constant use.

Heat gains from solar radiation may occur through glazed doors, although shutters or curtains may be required to improve privacy, security and heat loss characteristics during night-time or during periods of cold weather.

Doors also affect air flow patterns through building interiors, and therefore help control ventilation and cooling processes. Doors may be left open in order to encourage natural ventilation, particularly in hot climates, although this may create problems with privacy and security.

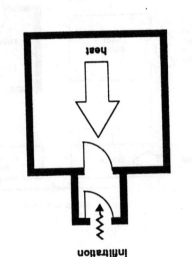

Above left
Section showing heat losses and infiltration through a door. Using thicker doors with good insulation properties and draught-sealing will reduce losses.

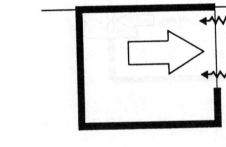

Above right
Plan showing a double set of doors for the main building entrance, reducing heat losses caused by infiltration and by people entering and leaving. The intermediate space provides a buffer zone of air between the main body of the building and outside. Using self-closing doors, locating the entrance away from prevailing winds or screening it with external landscaping will further reduce losses.

See also section six Architecture 'Plan' and section seven Building Elements 'Landscaping' and 'Ventilation'.

floors

Infiltration

Floors are the lower surfaces of rooms, located at ground level or at intermediate level between storeys. Ground level floors may be solid, in direct contact with the ground, or suspended, with an air space underneath. Floors are made of wood, concrete, stone, earth, or metal and can be covered with a range of finishes, including ceramics, fabrics, plastics and stone.

Heat losses through ground level floors depend on the thermal characteristics of the floor material, the size and edge condition, plus the soil type below solid floors (high water content increases heat losses) and the air temperature below suspended floors. In both cases, insulation placed underneath the floor or around the floor edge reduces heat losses. Insulation can be placed above the floor, but this prevents the floor thermal mass from being used to store direct solar gains. Suspended timber floors often need good ventilation underneath to protect them from decay.

For intermediate floors, heat transfer between rooms is resisted according to floor thickness and material. Heat losses occur through cracks and joints or through

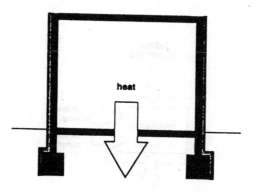

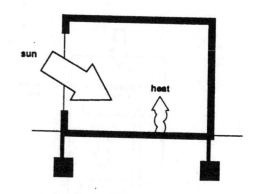

apertures such as trap-doors or stairwells. Heat gains may be stored in the floor thermal mass which absorbs solar radiation and re-emits it as heat. These floors are usually made of dark, heat absorbent materials. Heat gains also occur from solar radiation reflected by the floor into the building interior.

Services carrying hot water or hot air below the floor should also be insulated. Where desirable, these gains should be directed into room above and below, rather than absorbed by the floor.

Above left
Section showing heat loss through a solid floor slab into the ground. Insulation on the exterior of the wall below the ground will help prevent heat losses from spreading into the surrounding terrain, and may sometimes be added as a retro-fit solution.

Above right
Section showing heat gains from solar radiation which are reflected and absorbed by the floor thermal mass during the day and radiated as heat at night. The use of the floor in this way also reduces the temperature difference between the floor and the room, and therefore tends to increase comfort levels.

See also section six Architecture "Structure", section seven Building Elements "Windows", and section eight Solar Applications "Heating".

heating
sources

The heat source for space heating and water heating systems in buildings is often a boiler. Each boiler consists of fuel source, combustion area, and heat output by radiation and convection. Most use a conventional or balanced (on an external wall) flue to supply air for combustion and to remove combustion products.

Boiler types include gas-fired (quiet and quick response), oil-fired (vaporizing burner or pressure jet, relatively inefficient and needing stoking) and solid fuel (flue-less but rare). Back boilers are located behind open fires and stoves.

Condensing gas boilers use condensation to increase efficiency. Flue gases are cooled by an additional heat exchange area and water vapour condenses out, enabling reclamation of latent heat of the vapour. Because less fuel is consumed, and more is converted to heat, efficiency is greater and pollution is lower than conventional boilers. Efficiency is best at low operating temperatures, so condensing boilers are most suitable for low temperature hot water, warm water heating and large domestic hot water applications, but less so for buildings

with low heat loss, high incidental gains, intermittent use and short heating seasons. A condensing economizer is a heat exchanger added on to a conventional boiler.

Alternatively, an electrically-powered heat pump takes heat from external air, water or the ground and raises the temperature for heating. The vapour compression principle is the same as for refrigeration, and can be reversed to provide cooling. Using a gas or diesel engine to compress the refrigerant raises efficiency, but also raises maintenance and capital costs.

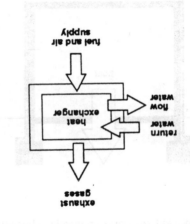

Above left
Non-condensing boiler. All except electric boilers require a flue to remove exhaust gases. Solid fuel and most oil boilers use a conventional flue, and draw air for combustion from the building interior. Gas and some oil boilers may use a balanced flue, and draw air for combustion from the building exterior.

Above right
In general, condensing boilers require a higher standard of construction than non-condensing boilers. For example, in order to reduce corrosion, flues for condensing boilers are normally aluminium or austenitic stainless steel.

See also section seven Building Elements "Cooling systems" and "Heating systems", and section four Solar Applications "Heating".

heating systems

Heating systems in buildings heat internal spaces and provide hot water. Specification depends on energy type, building shape, size and function, occupation pattern, structure, use of solar radiation, air patterns, expected heat losses and capital. Four arrangements are possible.

Isolated heaters heat specific areas of a building. Coal or wood-burning fires offer little control, and may be polluting. Storage heaters use electricity at night to heat a masonry core, emitting heat in the day, but offer poor control. Portable gas, kerosene and electric heaters provide regulated heat as required.

Central source systems are powered by gas, oil, solid fuel, or electricity. Heat is piped from a boiler to radiators filled with water or oil, or electricity may be used to heat elements in the floor or ceiling. User operated thermostats normally provide control. In forced-air systems, heated air is pumped into rooms by fans and ducts. A heat recovery system, using coil or plate exchangers, reclaims heat lost through exhausts.

Remote source systems provide heating and hot water for a number of buildings, sometimes as by-products

from a power station. Unless users are aware of energy costs, profligate use may occur.

Water heating may use the same boiler as for space heating. Water is heated and stored centrally, and dispersed upon demand. Alternatively, a water heater at the point of use supplies hot water as required.

Waste in heating systems is caused by inefficient heat generation, system losses and unnecessary use. Efficiency means providing heat and hot water only as required. Flexibility and user control are especially important.

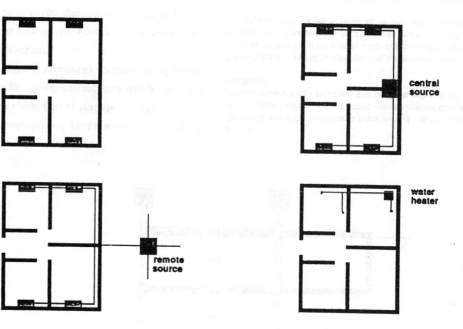

Top left
Plan showing rooms heated by a number of isolated heaters.

Top right
Plan showing rooms heated by a central source system.

Above left
Plan showing rooms heated by a remote source system.

Above right
Plan showing a water heater serving a total of three hot water dispensing points.

See also section seven Building Elements "Heating sources", and section eight Solar Applications "Heating".

insulation

Insulation materials are poor conductors of heat, applied to buildings to reduce energy lost through heat conduction and radiation.

Insulating the building envelope reduces heat losses to the outside. Walls may have external, internal or cavity insulation; windows may be double or triple glazed; and roofs, floors and doors may also be insulated. Insulation is applied to internal partitions which divide heated from non-heated spaces. The thickness and type of this insulation may be different from that for the rest of the building. Service elements, such as hot water pipes and tanks and boilers, should also be insulated to reduce heat losses.

Choice and location of insulation depends on availability and cost, the surface to be insulated, use patterns and climate. Efficiency of insulation depends on thickness, conductivity and workmanship. Possible materials include mineral wool and mats, cellulous fibres, polyurethane and urea-formaldehyde foams, and polystyrene beads and slabs. Insulation may be improvised, such as from polystyrene packing.

Insulation is one of the simplest and most cost-effective methods of energy-efficient design. Ideally, it is included at the design stage, although retrofit application is often possible, especially to service elements. Some insulation is nearly always better than none.

Careful design is necessary.

Uneven insulation of areas with low thermal resistance causes rapid heat transfer to the outside (thermal bridging). Condensation and damp can be a problem if insulation is incorrectly specified or installed.

Retro-fit insulation and multiple-glazing are extremely common methods of improving the energy efficiency of an existing building. If carried out, consideration must be given to the original building design, especially for water-proofing, ventilation and condensation.

See also section six Architecture 'Orientation', section seven Building Elements 'Ceilings', 'Doors', 'Floors', 'Shades and screens', 'Walls' and 'Windows', and section eight Solar Applications 'Cooling' and 'Heating'.

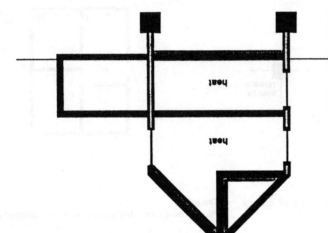

Above
Section showing the appropriate placement of insulation and use of double-glazed windows in a building with an internal garage and partially-exposed roof structure, and on a site with wet soil conditions.

landscaping

Landscaping modifies the effects of wind, solar radiation and precipitation on buildings. Energy-efficiency is therefore affected in three ways.

Heat loss. Wind passing over the surface of a building and wind-blown precipitation on the building fabric both conduct heat away, thereby increasing heat losses. Wind also disperses solar gains held in the external air, increasing differences between interior and exterior temperatures, and so encouraging heat loss.

External walls, hedges and trees in the landscape around the building reduce wind speeds and exposure levels, therefore reducing heat losses. However, positioning must be carefully undertaken as the creation of wind funnels will have an adverse effect. Where through access is needed, walls can have openings small enough as to minimize reductions to wind protection.

Shading. Overheating in buildings may occur, particularly in summer conditions on the south side of a building with a large area of south-facing windows. Trees and tall shrubbery provide shade, especially when placed in rows, and thus reduce unwanted heat gains. Trees

require careful selection to provide the right combination of shade and shelter. For example, the high foliage of palm trees provides shelter from high level winds and shade from high angle summer sunlight, without obstructing low angle winter sunlight; deciduous trees lose their leaves in winter and thus allow the sun to pass through when it is most useful.

Ventilation. Careful positioning of external walls, hedges and trees reduces cooling requirements by controlling ventilation and directing air movement.

sun

calm air

wind

Above

Site section showing the use of landscaping to provide shading from solar radiation and shelter from winds and airborne particles. Trees take a number of years to reach maturity, and may block views and disrupt foundations. It may therefore be more feasible to adapt site designs to existing landscaping patterns. Artificial features such as walls may also be incorporated. Controlling wind speeds is most important during where or when winds are strongest and heat losses are greatest. Shelter is best provided by intermittent planting rather than by a continuous barrier. The effect of buildings on the micro-climate must also be considered.

See also section four Environment "Environmental concerns", section six Architecture "Site", "Orientation" and "Plan", and section seven Building Elements "Shades and screens".

lighting

Light is the visible part of the electromagnetic spectrum. Lighting is the provision of artificial light, supplementary to daylight, for the use, comfort and safety of building occupants. Three basic sources of artificial light are used in buildings: burning fuels (such as gas, candle wax and wood), electric fluorescent lights, and electric incandescent lights.

All light sources differ widely in efficiency. Nearly all of the energy consumed is ultimately converted into heat gains in the room. Although these gains are often beneficial, lighting can create uncomfortable temperature levels, particularly in intensively or closely lit areas, thereby creating energy demands for cooling.

Careful design increases efficiency. Efficient light sources and fittings should be selected and maximum advantage should be taken of daylight while meeting the specific requirements of users. Lighting does not have to be uniform across a room, and can be provided only where needed. Flexible arrangements, such as those including movable task lights, track-lights and uplighters, can allow for changing use patterns to be accommodated.

Lighting control is important for energy-efficient design. Users must be aware of their responsibilities, and be able to make changes as required. Where properly used, automatic and semi-automatic controls can also be effective. These include time switches operating to a pre-set schedule, photo-electric switches reacting to daylight conditions, and occupancy-linked controls sensing the presence of users. Detailed attention to the function and location of components is essential if the controls are to be used effectively.

Rooms designed with small windows will create increased energy demands for artificial lighting. Illuminance levels should be appropriate for the room function, meeting specific task and general needs. Artificial lighting controls can be linked to natural light levels to minimize energy demands.

See also section seven Building Elements "Roofs", "Shades and screens", "Sunspaces" and "Windows".

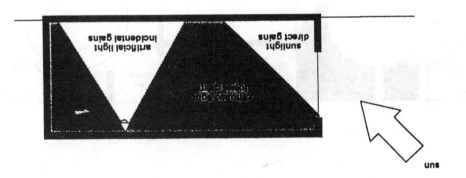

sun

sunlight direct gains

artificial light incidental gains

Above
Natural light provides sunlight and direct heat gains to the building interior, plus diffused daylight and lower heat gains away from the window. Artificial lighting systems, often necessary even during the day for locations away from windows and to provide task lighting, offer control over light conditions and add incidental heat gains.

roofs

Roofs are the uppermost surfaces of buildings, giving protection against precipitation, winds and heat exchange between building interior and exterior. Roofs can be made of a range of materials, including wood, metal or concrete for structure and wood, metal, clay, plastic, thatch or vegetation for protective covering. Roof condition affects both the overall condition and the energy efficiency of the building.

Heat losses occur by conduction through the roof fabric and by convection through gaps in the fabric and eaves. Where undesirable, these losses are controlled by adding insulation above the ceiling, or next to the roof if the roof space is used. For flat roofs, insulation may be placed inside ("cold roof") or outside ("warm" or "inverted roof") the roof structure. Insulation capacity depends on the thermal properties of roofing materials, size of roof space and insulation type. Over-insulating at the bottom of the roof space means that any water pipes or tanks in the roof space can freeze during cold weather.

Ventilation above the insulation layer is essential to avoid condensation. In hot climates, ventilation should be increased to reduce energy demands for cooling.

Heat gains occur through solar radiation absorbed by the roof fabric and radiated as heat to the roof space or rooms below. Gains are greatest with dark coloured pitched roofs orientated toward the sun. Glazed roofs with no intermediate ceiling provide direct heat gains and light to rooms below. In hot climates, a massive roof can delay the ingress of solar heat gains until the evening and lose heat to the sky overnight. Performance can be enhanced by using movable insulation or roof ponds.

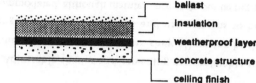

- ballast
- insulation
- weatherproof layer
- concrete structure
- ceiling finish

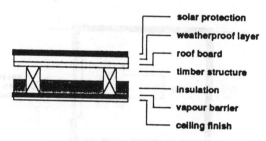

- solar protection
- weatherproof layer
- roof board
- timber structure
- insulation
- vapour barrier
- ceiling finish

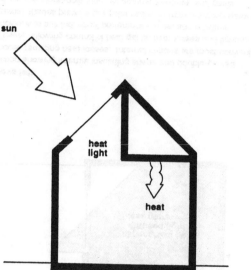

sun

heat
light

heat

Top left
Section through a typical inverted flat roof. Ballast may be gravel, paving slabs or other heavy material.

Above left
Section through a typical cold flat roof. Good ventilation between insulation and roof board reduces condensation risk in the winter.

Above right
Section showing direct heat gains through the roof glazing and indirect heat gains radiated from the roof and roof space.

See also section six Architecture "Structure", section seven Building Elements "Ceilings", and section eight Solar Applications "Cooling" and "Heating".

shades and screens

Shades and screens modify the performance of windows in buildings by preventing excessive day-time heat gains and solar glare. They are particularly useful on large south-facing windows during summer conditions. Shades and screens also reduce heat losses, particularly at night, in the winter months, and on any window where the admission of light is not constantly necessary for the use of the room within.

Many different types of shades and screens exist, including horizontal projections and eaves, canopies and awnings, curtains and blinds, shutters, louvred or slatted blinds or fins, and fixed or retractable screens. Most can be designed into new buildings or fitted retrospectively to existing ones. In general, shades and screens can be divided into three types: *external* shades absorb solar radiation before it strikes the window, and control heat gains; *intermediate* blinds may be placed in the cavity of double or triple glazed windows; *internal* shades intercept solar radiation after it has passed through the window to control heat gain, glare and illumination levels, and also insulate the glazing to control heat losses.

Adjustable shades and screening are preferable to permanently fixed devices as they allow for both changing weather conditions and use patterns to be accommodated, although maintenance costs can be high. Most types can be quickly operated to control heat gains and heat losses. This can be done manually or by automatic control motor-driven mechanisms responding to a pre-set timing schedule or prevailing weather conditions. User over-ride controls for automated shades are highly desirable.

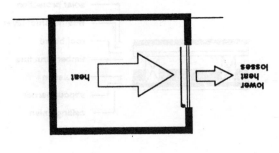

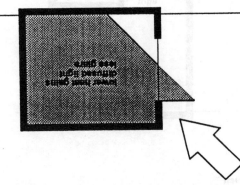

Above left
Section showing internal insulating shade and double-glazed window, reducing heat losses. Internal shades are most convenient for users, allowing control of heat gains, heat losses and lighting conditions to suit individual preferences. Internal shades prevent any solar radiation from reaching windows, and are therefore most effective against day-time heat gains.

Above right
Section showing external horizontal shade reducing heat gains from solar radiation. Fixed external shades may not be adequate for low altitude winter sun and changing weather conditions.

See also section three 'Human Factors 'Control' and section seven Building Elements 'Landscaping', 'Sunspaces' and 'Windows'.

sunspaces

An atrium or a conservatory, collectively called sunspaces, consists of a glazed enclosure functionally separate from the rest of the building. An atrium is contained within the building envelope, usually in the middle. A conservatory is added to one side of the building, usually south-facing.

Sunspaces mediate between the internal space of the building and the outside air, modifying heating, lighting and ventilation conditions. As with windows, the glazing reflects or absorbs some of the radiation from the sun. The remainder passes through to be absorbed by the internal surfaces and contents, and is released back into the room.

Sunspaces can be used as unheated direct gain areas acting as an extension of the dwelling area, or as collector areas, from which warm air is extracted to remote storage, often using fans. They may include a masonry or water thermal storage wall which absorbs and re-emits heat to the building.

Large temperature swings may be experienced during the day and, because of the large glazing areas in

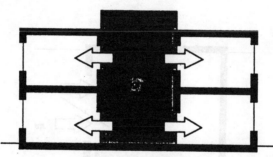

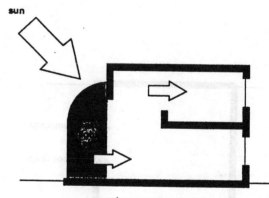

sunspaces, these occur more rapidly than in normal rooms. Shading and ventilation are therefore essential to prevent over-heating during the day and insulation should be considered to avoid extreme cooling at night.

Sunspaces can be part of new buildings or added onto existing buildings, although the latter is more common with conservatories than with atria. Although sunspaces may provide small energy savings in relation to their cost, they often make the building more attractive to use.

Above left
Section through a building with centrally-placed atrium. Modified heat, light and air are dispersed outward to the rest of the building. Atria may be used as a retro-fit addition over a central courtyard.

Above right
Section through a building with conservatory on the south side. Modified heat, light and air are dispersed into the rest of the building. Conservatories are most common in domestic buildings, where they create additional low-cost space for occasional recreational use rather than for continuous occupation.

See also section seven Building Elements "Shades and screens" and "Windows", and section eight Solar Applications "Heating".

ventilation

Ventilation is the changing of air in buildings to control oxygen, heat, odour and contaminants. Ventilation may occur in four forms.

Natural ventilation. Air movement is caused by pressure or temperature differences across building apertures. Building orientation, form, plan and user actions also alter air flow paths. Natural ventilation consumes no energy and has few if any running costs, but depends on weather conditions and can be difficult to control, often creating draughts or stuffiness.

Mechanical and air-conditioned ventilation. These are energy-driven alternatives to natural ventilation, normally dictated by building type, site and function. They can be particularly efficient as supplements to natural ventilation. Mechanical ventilation uses fans and ducts to supply and extract air in localized areas such as a kitchen. Air conditioning both treats and supplies air. It is particularly useful to cool air below ambient temperatures.

Mechanical and air conditioned systems consume energy (usually electricity) for air movement, filtration,

heating, cooling and humidification. Indiscriminate design, operation and maintenance lead to very high energy consumption: in particular, heating and cooling of the same air should be avoided.

Infiltration. Infiltration is the unintended leakage of air through imperfections such as cracks around doors and windows, and is closely related to the quality of building construction and maintenance. Although beneficial to ventilation, infiltration raises energy demands.

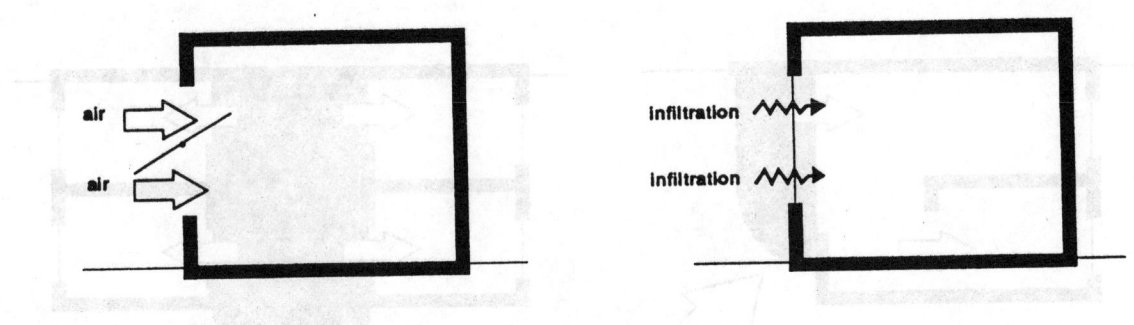

Above left
Section showing natural ventilation through an open window. Natural cross-ventilation may be encouraged by opening windows on opposite sides of the room. Users may interfere with predicted ventilation patterns by closing windows and blocking air vents.

Above right
Section showing infiltration through cracks in a window frame. Like natural ventilation, infiltration is dependent upon meteorological conditions and so can exceed, satisfy or fail fresh air requirements. Some infiltration must take place to reduce the risk of condensation.

See also section seven Building Elements "Cooling systems", "Doors", "Insulation" and "Windows", and section eight Solar Applications "Cooling".

walls

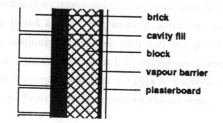

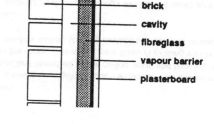

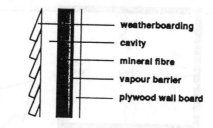

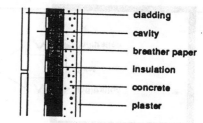

Walls provide shelter, security and structural support in buildings. Load-bearing walls are usually made of clay bricks or stone, although concrete, wood, metal, earth, straw and plaster are also used. Non-loadbearing walls, such as internal partitions and external cladding, are usually lighter in construction. Walls are important for energy-efficient design in three ways.

Insulation. Internal wall insulation lowers the thermal mass of external walls and therefore reduces the energy required to heat up the building, but can also lead to overheating. For constantly heated buildings, externally insulated walls reduce temperature fluctuations, although the building may take longer to heat up. A wall cavity can be partly or completely filled with insulation, with provision for ventilation and drainage as necessary.

Thermal storage. Solar radiation is absorbed through the external wall, conducted to the inner surface, and radiated as heat to the interior. Thermal mass efficiency is set by wall thickness and material, which determine storage capacity, and by surface colour, which determines

absorption. In general, thicker walls help stabilize internal temperatures.

Space definition. Internal walls define the boundaries of rooms. Where there are temperature differences between rooms, heat transfer is resisted according to wall thickness and construction material. Internal walls affect air flow, and therefore modify ventilation patterns, heat convection and condensation. Whether these processes are beneficial or not depends upon room function and individual user preferences.

Top left
Insulated wall with brick, block and filled cavity.

Top right
Insulated wall with brick, plasterboard and partly-filled cavity.

Above left
Insulated wall with plywood and boarding.

Above right
Insulated wall with concrete and cladding.

See also section six Architecture "Structure", section seven Building Elements "Floors", "Insulation" and "Shades and screens", and section four Solar Applications "Heating".

windows

Windows provide light, ventilation and view. They are normally glazed with a single or multiple layer of glass. Low-emissivity glass provides triple-glazed performance from two sheets of glass. Photochromic and electrochromic glass, which modify their transmission characteristics according to light conditions, may become more common in the future.

Windows contribute to the heating of buildings. Some solar radiation is reflected back or absorbed by the glass. The rest passes through to be absorbed by the room surfaces and contents, and is released back into the room. This heat gain is useful when it reduces energy demands for heating; windows should be placed to meet functions within.

Windows are also a major source of heat loss. If the temperature is higher inside than outside, heat passes out through the window and lowers the interior temperature. Although openable windows provide environmental control, this may cause heat loss, as may air infiltration through cracks around window frames. Heat losses are greatest in the winter.

Solar gains should be balanced against heat losses by adjusting window size, type, number and placement. In general, larger south-facing windows maximize heat gains, although night-time heat loss will be equal to gains. North-facing windows minimize summer solar gain without shading may increase energy demand for cooling; smaller north-facing windows minimize heat losses, and should have the smallest area compatible with acceptable daylight and view. Multiple-glazing reduces heat loss by providing an insulating cavity between glass layers.

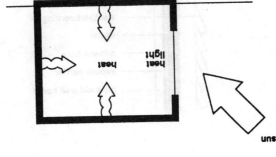

Above left
Section showing heat gains through windows. Direct solar radiation is reflected and absorbed by the ceiling, walls and floor and radiated back into the room. Gains and glare through existing windows may be controlled by adding shading devices and external landscaping.

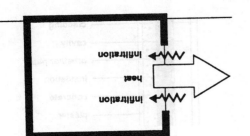

Above right
Section showing heat losses through windows. Losses and ventilation through existing windows may be controlled by adding internal screens and secondary glazing, by sealing cracks, and by modifying internal air movement patterns.

See also section seven Building Elements "Insulation", "Landscaping", "Lighting", "Shades and screens", "Sunspaces" and "Ventilation", and section eight Solar Applications "Heating".

solar applications

Solar applications use the building form and fabric to admit, store, distribute and re-emit solar energy by means of radiation, conduction and natural convection.

For heating, passive systems obtain solar energy from glazing (direct gain), thermal mass (indirect gain from solar energy, first absorbed by the thermal mass and then radiated or convected as heat), or a separate storage system (isolated gain, usually heat transferred from a thermal mass away from the main building). Active solar systems work on similar principles but require additional energy inputs to transport the heat from the collection system to where it is needed.

For cooling, passive systems utilize thermal mass (thermal mass radiates heat out of the building at night), the ground (heat is lost into the ground below), ventilation (solar energy is used to encourage air movement) and evaporation (latent heat reduces air temperatures). Active systems may make use of fans to increase ventilation and evaporation rates and to expel heated air to the outside.

As described in *section seven Building Elements*, many of these principles are already included in the design process as appropriate to building elements and the building as a whole. The system arrangements and devices described in this section, *Solar Applications*, are designed largely with solar heating and cooling considerations in mind. Solar systems try to maximize passive and active benefits and to apply them, in the form of design solutions, to buildings.

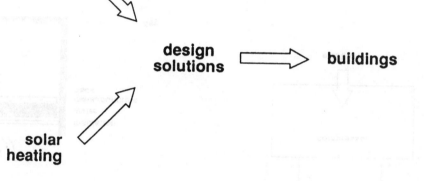

Above
The application of solar heating and cooling as design solutions to buildings.

cooling

Passive and active cooling systems in buildings help cool internal spaces. Four arrangements are possible.

Night radiation. A thermal mass will normally cool on clear nights by radiation and provide cooling by day to the building interior by radiation and convection. The effects are maximized by exposing a thermal mass (such as a wall, roof or water-filled roof pond) at night and insulating it from the exterior during the day. These systems require the use of movable insulation.

Evaporation. Water (such as a pond, roof pond or sprayed water) will increase the humidity of hot and dry air but reduce the dry bulb temperature, thus increasing comfort in hot, dry weather.

Ventilation. Solar radiation may be used to heat a "thermal chimney" and so create cooling by convection. Warmed air rises up and out of apertures placed high in the building, while drawing in cooler external air through apertures placed low down in the building and creating air movement to increase cooling. Wind scoops directing external air movements into the building can also create cooling by convection.

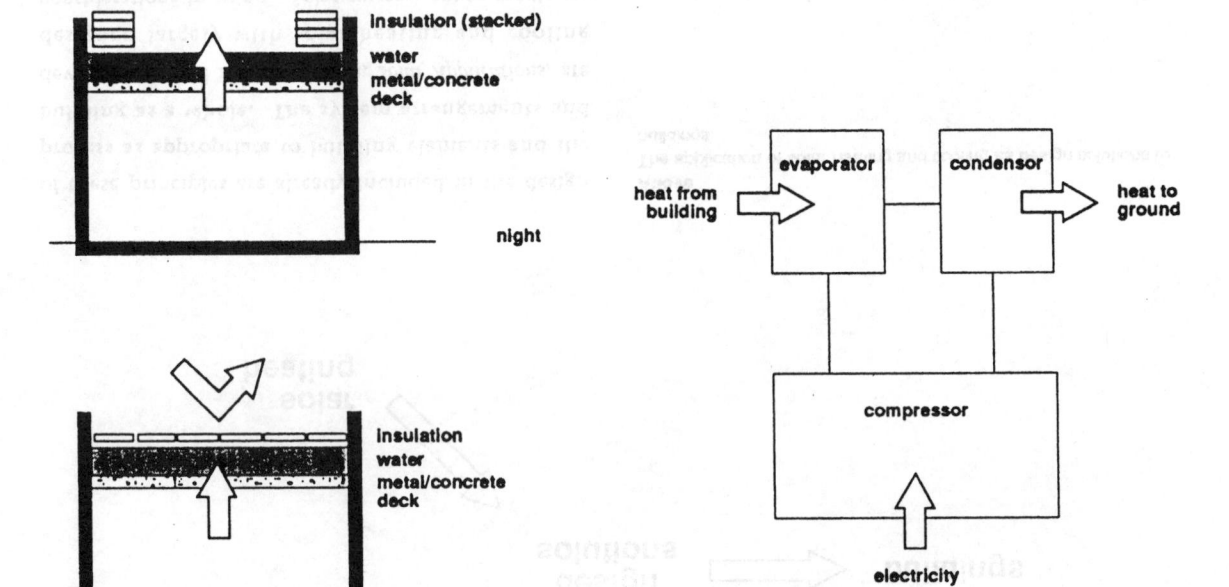

insulation (stacked)
water
metal/concrete deck

night

insulation
water
metal/concrete deck

day

heat from building — evaporator — condensor — heat to ground

compressor

electricity

Ground temperature. The difference between the internal building temperature and the relatively constant external ground temperature may provide cooling: internal heat is conducted into the ground as a beneficial heat loss.

Top left and above left
Sections showing cooling by the use of a thermal storage roof pond with movable insulation. A metal or thin concrete deck assists heat radiation and forms the room ceiling below. During the winter, the procedure may be reversed to allow day-time collection of solar radiation and night-time radiation of heat into the building.

Above right
Vapour compression heat pump with an electric compressor arranged for cooling. During the winter the process may be reversed for heating.

See also section seven Building Elements "Cooling systems", "Heating sources" and "Ventilation", and section eight Solar Applications "Heating".

heating

Passive and active systems in buildings heat internal spaces and contribute to hot water provision. Three arrangements are possible.

Direct gains. South-facing solar walls introduce large amounts of heat and light directly into the building interior. However, they can overheat the interior in the summer, increase heat losses in the winter, and suffer from problems with condensation and glare. Other design arrangements to control direct gains are covered in *section three Building Elements* particularly "Shades and screens", "Sunspaces" and "Windows".

Thermal mass. Vertical south-facing devices such as thermal mass and Trombe walls (masonry or concrete construction with external surface glazing) and water walls (water-filled construction with external surface glazing, offering faster heat transfer), and tilted or horizontal devices such as thermal mass floors, roof ponds (water mass located in the roof space) and black attics (dark concrete or masonry slab located in the roof space) all absorb, conduct and then radiate and convect solar radiation as heat to the building interior. Heat

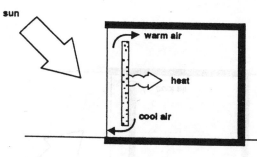

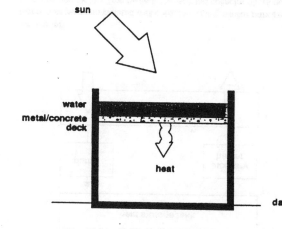

distribution may be augmented by vents or fans. Movable insulation may be used to minimize night-time heat losses or to reverse the heat transfer process, providing night-time or summer cooling as required. Remote storage walls are insulated to minimize heat losses, with heat transferred to the main building interior by convection, often fan-assisted.

Passive solar collectors. Roof space collectors use an inclined, glazed and south-facing roof, darkly coloured

continued

Above left
Section showing a Trombe wall. The glazed mass wall, usually south-facing, is warmed by solar radiation and radiates and convects heat from its inner surface to the interior. Vents at top and bottom control convection heat gains from air warmed between the wall and glazing, and prevent night-time heat losses. A glazed mass wall operates on the same principle, but has no air vents. A Trombe wall is particularly useful in mountain climates with cold air and clear skies.

Above right
Section showing a roof pond. The pond, covering at least 50% of the ceiling area, provides uniform heat radiation to rooms below. Insulation can be placed over the pond to prevent night-time heat losses, or alternatively during the day for summer cooling.

and insulated on the inside where unglazed, to pre-heat and convect air in the roof space to the building interior. They may also be used in conjunction with a roof pond or black attic. Isolated wall collectors use a lightweight glazed collector attached vertically to the south wall of a building. Air trapped between the collector and wall is heated and and distributed by thermo-siphoning (it expands and rises) to the building interior via connecting apertures, while cool air from the interior is drawn in at the base of the collector. Convective loop solar collectors use a similar principle to heat up an air mass positioned below the building interior and convect it upward into internal spaces or to a thermal storage mass, while cool air from the building interior is either expelled or drawn back to the collector.

Active solar collectors. These are usually added on to the building without forming part of its structure, and require additional mechanical devices to move heat gains. Active collectors can be either flat-plate (a network of pipes heated behind a glazed cover and insulated by a rear black panel) or concentrating (reflective materials focus solar radiation onto a small area). From these collectors, pumps and fans transfer solar energy as heated air, water or other liquid to boilers and heat exchangers, and then to the building interior as hot water and warm air. Air-based systems are preferred where only space heating is supplied, and water-based systems are preferred for combinations of space heating and cooling and hot water provision.

Active solar collectors may also provide electricity, either directly, using photovoltaic cells to convert solar radiation, or indirectly, using concentrating collectors to heat liquids in order to power a generator.

In all active solar arrangements, overall system efficiency may differ significantly from collector efficiency. Active systems often require an auxiliary or back-up conventional heating system for use during inappropriate solar conditions, and are normally reserved for use at low latitudes.

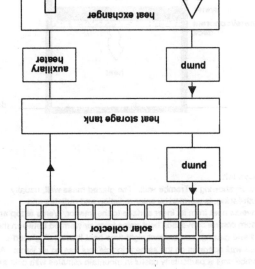

Above left

Section showing an isolated wall collector. Using similar principles to the vented air in a Trombe wall, the isolated collector offers quick response air heating, and can be used as a retrofit addition.

Above right

Typical arrangement of a flat-plate active solar collector and system. There are two heat loops: collector to storage, and storage to rooms via a heat exchanger. The collector is usually located on a south-facing inclined roof. Operation usually requires little user attention, but system complexity can make it prone to failure.

See also section seven Building Elements "Cooling systems", "Cooling sources", "Heating systems", "Roofs", "Sunspaces" and "Windows", and section eight Solar Applications "Cooling".

part three

monitoring

monitoring systems

Energy-efficient designs are often monitored to evaluate performance-in-use. This tests the building against the predicted performance and gives designers objective information on costs and benefits of installed systems. In general, systematic monitoring will increase the quality and performance of buildings and reduce the time between design implementation and understanding its consequences. Monitoring will become more widespread for energy-efficient designs, for buildings in general and for the external environment.

Detailed monitoring means gathering data by recording instruments or data-loggers for interior and exterior temperatures, solar radiation, humidity, wind speed and direction, the opening of doors and windows, hot water use and air change, and from electricity, gas and hot water meters. Samples are taken at regular intervals during the day, such as every 30 minutes, and recorded automatically on a data logger controlled by computer or calculator programme. Monitoring should last for at least one year of normal occupancy and use. In some cases mechanical loggers are used and the data

transcribed by hand. In other cases data are sent automatically by modem or radio transmitter to a remote computer. Time series data from the logger are transferred to tape or disc and analyzed by computer. Results may then be read as graphical or tabular output.

Remote recorders can be problematic in use. Computers can suffer from power surges, human tampering and mechanical failures. Correct location of sensors is important to prevent misrepresentation of building performance and to prevent damage to sensors.

design knowledge

↑

monitoring

↙   ↘

quality   performance

Above

Three benefits of using monitoring systems in buildings designed from the perspective of energy efficiency.

Simple monitoring using regular meter readings, spot checks and a professional survey can often give useful information.

monitoring by example

This guide has demonstrated the factors which designers should consider when designing from the perspective of energy efficiency. Many examples of energy-efficient buildings exist throughout the world, and some of the best have not been designed by architects or engineers: they are traditional, vernacular designs which respect local environmental and ecological systems without making undue demands on energy resources. Such buildings can be found in most countries, using established construction techniques and thereby transferring design knowledge from one generation to the next.

In modern societies this transfer is being lost, so that knowledge about building techniques is no longer passed on automatically but by formal education and training and by systematic monitoring and feedback. Energy-efficient design is pioneering this kind of knowledge transfer in architecture. Monitoring is crucial to shorten feedback periods and to give designers factual knowledge about the benefits and costs of buildings.

Each of the countries briefly described in this section have made a commitment to the idea and practice of energy-efficient design. Some, such as Yugoslavia and Hungary, already have sophisticated and advanced programmes; others, like Czechoslovakia, Poland, Malta, Cyprus, Turkey, and Bulgaria, have experimental buildings but no installed base; some, such as Albania, are looking to learn from the experience of others before they apply these techniques across a range of buildings. In all cases there is huge potential for effective energy-saving carried out in a socially and environmentally responsible way.

Above
Three ways of passing on knowledge about energy-efficient design.

energy efficient design

traditional design

education and training

monitoring and feedback

64

albania

Albania is self-sufficient in fuel but has low levels of technological and economic development. It is therefore largely dependent upon imports of foreign materials and equipment. Technical and design expertise for heating and ventilation systems in buildings are being developed to counter some of these problems.

The country experiences a mixture of Mediterranean, hilly and mountainous climates that produce a varied heating season of between 4 and 6 months in the year. Because the climate varies between regions, different energy-efficient design solutions have to be developed and applied to each of them. Most dwellings are heated by wood, coal or gas-oil originating from within Albania, with district heating systems also being used in some areas. Clay and hollow bricks are the most common kind of building material for traditional buildings throughout the country.

The main approach to energy efficiency in Albania is designing for site, orientation, shading from solar radiation and sheltering from the wind rather than the use of advanced technological systems. Draft town

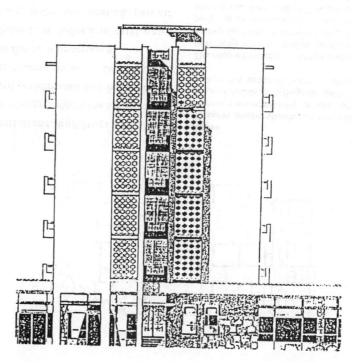

Above
Indicative graphic showing an energy-efficient apartment block in Tirana. Apart from its energy-efficient features, the block is typical of others throughout Albania.

planning codes have been introduced to encourage these design practices.

Although no examples of energy-efficient solar buildings yet exist, an energy-efficient apartment block has been built in the centre of Tirana. The block has a system-built panel construction and faces south to avoid over-shadowing and to maximize winter solar gain. The flats are heated by wood-burning stoves and are clustered in groups of 3 to 4 around a central staircase to reduce heat losses.

bulgaria

Bulgaria has developed a pragmatic approach to energy-efficient design by maximizing existing resources. Energy-efficient design features have been included in several buildings, including schools and hotels. Although few energy-efficient design projects yet exist, and hence there is relatively little practical experience, interest in energy-efficient design is growing.

The heating season lasts for 5 to 6 months with an additional requirement for cooling for part of the year,. Most of the energy consumed is imported solid and liquid fuels. Space heating accounts for 75% of energy consumed in housing. Certain sectors such as shops and offices require air conditioning in the summer months. It is estimated that by the year 2000 air conditioning will use 3-4% of electricity consumed in the summer.

Bulgaria also contends with construction methods and materials which result in poor building energy efficiency. New apartments are often of panel construction, leading to problems with a lack of insulation materials, low U-values in existing panels, poor sealing materials and lack of flexibility for retrofit.

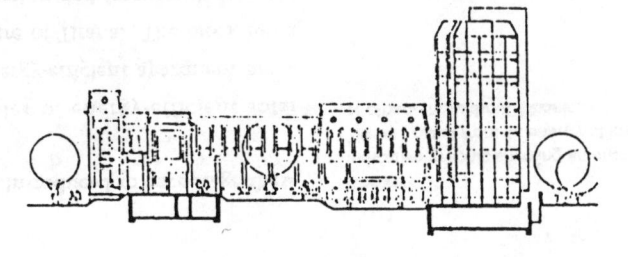

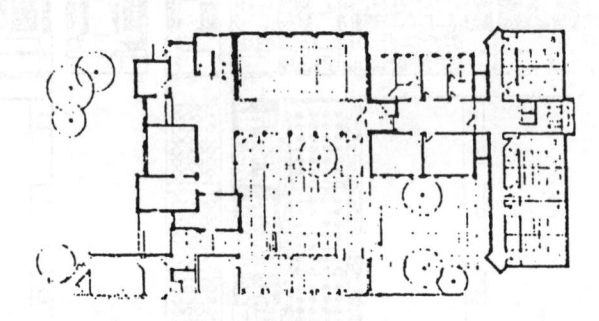

Despite these limitations, buildings have been designed with lower energy requirements through improved fabric thermal performance and better use of underfloor and district heating systems. Designing for site and orientation has also been undertaken.

A new hotel in Bulgaria includes a mineral water heat source, heat pumps, under-floor heating, part air conditioning and part forced air. It also uses external insulation of 4 cm mineral wool and a curtain wall of reflective glass to control heat gains on the south side.

Above
Indicative graphic showing the south elevation and plan of the proposed residential wing, or hotel, for the Central Home of the Union of Architects in Bulgaria (UAB). The wing, on the right, was designed by architects Guergui Hadjivanov and Ilia Damianov.

The wing will incorporate demonstration use of advanced technology including heating, ventilation, air conditioning and lighting equipment and their controls, monitoring systems, and high performance thermal insulation and cladding. It is intended that training and other advice will be given for residential and non-residential architects and building professionals.

cyprus

Cyprus imports most of its energy, which, along with hot summers and mild winters, offers great incentives and potential for energy-efficient design. The heating season lasts 4 months and diurnal temperature ranges reach 16° C in the summer and 10° C in the winter.

Traditional dwelling construction uses sun-dried 40 cm walls and roofs of woven crushed cane mats placed over wooden rafters and covered by roof tiles. Modern construction often uses a loadbearing reinforced concrete structural frame with 20 cm hollow brick infill walls and horizontal concrete slab roofs. Windows and doors usually have wooden shutters. Expanded polystyrene, polyurethane foam and perlite insulation materials are manufactured in Cyprus. Rock for making rock wool is also available.

Energy-efficient design has to contend with a variety of heating and cooling technologies which do not depend on single energy sources. Typical energy-efficient measures include external cavity walls, shading devices, lightweight concrete in roof voids and double-glazed windows. Almost 90% of houses and 50% of hotels use

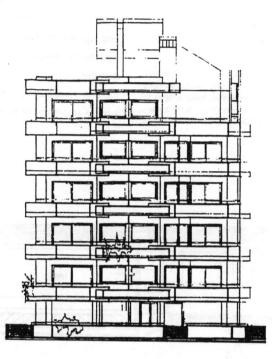

solar water heaters for hot water.

No regulations for thermal insulation yet exist but official tables for the thermal transmittance of materials and instructions for site, orientation, form, insulation and shades have been issued. Adopting recommended coefficients of 1.1 W/[m²K] for external walls and 1.0 W/[m²K] for roofs is expected to produce 50% reductions in energy consumption for heating and cooling in dwellings while maintaining optimum comfort conditions.

Above
Indicative graphic showing the south elevation of the Frangou apartment block in Nicosia. The block was designed by architects G. Mavrommatis & Associates in 1986 and completed in 1988. It incorporates a combination of active solar collector panels with an oil-fired heating system. Automatic time controls in each unit can be over-ridden by the occupants as required.

czechoslovakia

The Czech and Slovak Federal Republic has several energy-efficient design projects involving a number of different building types. Many of the projects are experimental and use simple and readily available technology and building materials as a practical response to capital and materials shortages. This has the advantage of keeping technical complexity and costs to a minimum. Amongst the most prominent buildings are the following.

An apartment block in Valtice near Breclav in Moravia contains nine flats. It has passive solar glazed loggias on the south side, solar collectors, window recuperators, thermo-reflecting paper-hangings and high insulation standards. One apartment consumes 6.38 MWh/year, about 66% of the average heat consumption for a unit of this kind.

A cultural centre in Ceská Lípa demonstrates some of the problems facing energy-efficient design in Czechoslovakia. A 14 year gap between outline design in 1976 and expected completion in 1990 had to be endured due to a lack of finance and poor availability of

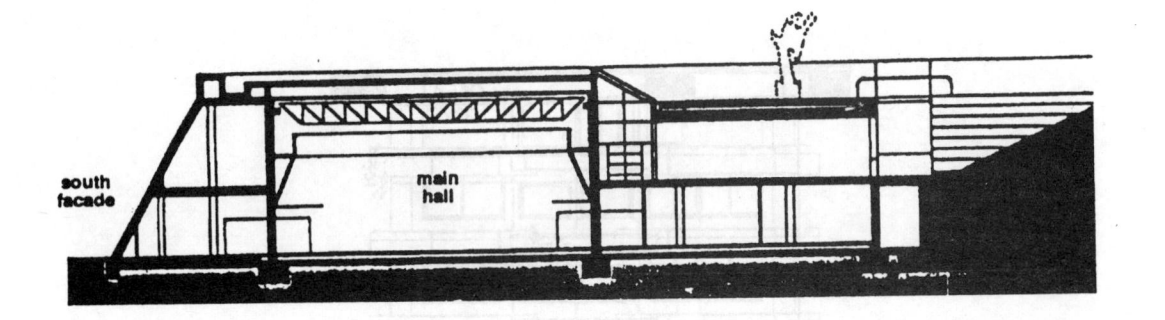

south facade   main hall

materials.

The centre uses a single-glazed air-collector on the south side and a thermally-insulated concrete slab floor with ducting under the main hall. The air collector warms the main hall, with the concrete slab storing heat and re-emitting it during the winter and cooling the hall in the summer. The ventilation system links the solar collector and the floor and is manually controlled. Conventional heating is installed in the smaller halls, meeting rooms and offices in the rest of the building.

Above
Indicative graphic showing a section through the cultural centre in Ceská Lípa. The building is positioned partly underground in order to reduce heat losses and to preserve the view from an adjacent historic monastery. An energy saving of 25% compared with a conventional building of the same size is expected.

hungary

Hungary has an established architectural and research tradition in bio-climatic design producing many examples of energy-efficient buildings, including individual dwellings, apartment blocks, and offices. Amongst the most prominent are the following.

The passive solar house at Pécs is situated on a south-facing slope. The design uses a two-storey glazed Trombe wall with motor-driven insulating shades on the south side and a lower north side. Other energy-efficient features include compact form, double-glazed windows, extensive draught-stripping, a roof garden, a partially embedded atrium with floor used for heat storage, and an additional gravel heat store under the staircase. An auxiliary gas-fired heating system supplements the solar space heating.

Another recent initiative is the passive solar housing community to be built at Kalocsa. Energy-efficient features are to include the following: east-west axis of streets and roof ridges, grouping of houses in rows, and L-shaped buildings with sunspaces; buffer zones located within building plans, carefully designed internal air flows; highly-insulated elements on the buildings' north sides, mass walls with transparent insulation and shaded glazing on the south sides, double party walls of heavy blocks, ventilated sloping roofs, roof gardens; warm water floor heating, and high levels of comfort, control and security.

As with other Hungarian work, the Kalocsa project represents advanced monitoring, feedback and simulation, an essential part of future energy-efficient design.

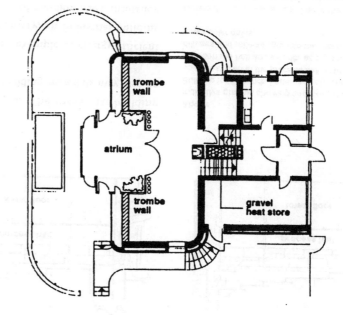

Above
Indicative graphic showing the ground floor plan of the passive solar house at Pécs. The house was designed by the architect J. Szász and the engineer L. Fülop in 1982-83 and constructed in 1984-85. The site was provided by the local council, who also financed most of the building costs.

A family is in residence, but the house is still in the experimental stage of operation. So far it has proven to be relatively slow in responding to changing weather conditions and user needs. It is intended that a user manual will be produced in the future. Microprocessor controls and the maintenance required for them are not considered feasible.

malta

Malta depends entirely on imports for fuels, leaving it vulnerable to price rises and fluctuations in supply. Over 40% of the total energy demand is consumed in buildings, of which 60% is for domestic space heating. At present, energy consumption by air conditioning for cooling is low.

Recent changes in construction methods and building materials have made many buildings less energy-efficient. New buildings use thinner walls and larger windows, and tend not to use courtyards which could stabilize internal temperatures. Although rarely used, polystyrene sheets are the most common insulation. No building regulations regarding energy losses yet exist.

Conversely, there is considerable potential for energy conservation, particularly because of the climate consisting of hot summers, mild winters and large diurnal temperature ranges.

All of these factors have lead to a great deal of interest in energy-efficient design. In the short term, energy-intensive space heating and water heating offer

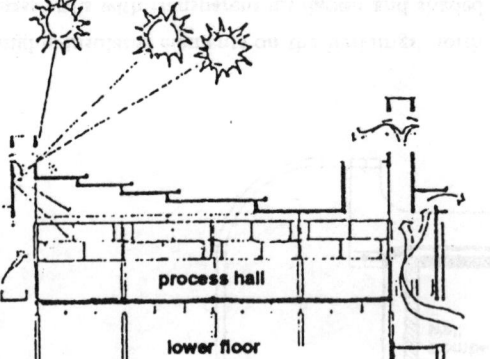
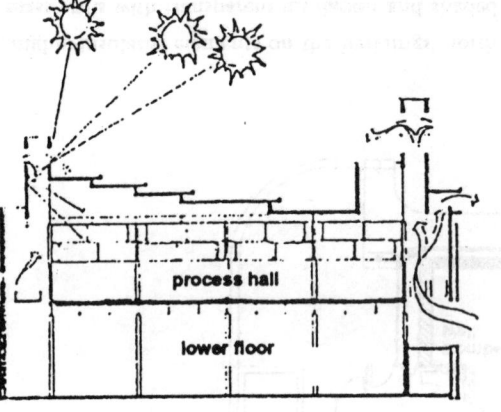

the greatest opportunities. The vernacular building tradition and available building materials may also be exploited.

The most prominent example of energy-efficient design is the Farsons brewery, an industrial building which uses a combination of building mass, insulation and air flows to provide summer passive cooling without using air conditioning. Summer night ventilation keeps temperatures in the process hall below 27° C for all but the hottest of August days.

Above
Indicative graphic showing sections through the Brewery Process Building for Simonds Farsons Cisk. The brewery design minimizes solar heat gain in the process hall by using the circulation space between the hall and exterior as a glazed insulating layer which diffuses and reflects light into the process hall and thus prevents direct solar gains.

During the day, heat is absorbed in the concrete and masonry thermal mass (left). At night, heat is radiated into the atmosphere by ventilation through vents positioned high up in the building (right). In the summer, these vents open and close automatically according to the time of day.

poland

Although a net exporter of energy as coal, Poland imports much of the energy it consumes as oil and natural gas. Increased energy-efficiency would reduce expenditure and improve the environmental condition of the country as a whole.

Heating is required for a large portion of the year: space heating accounts for 62% of all domestic energy consumption, mainly as combined heat and power and as district heating: most apartments (which make up 60% of dwellings) are connected to district heating networks. These apartment blocks tend to be constructed of concrete or brick. The most commonly used insulation materials include mineral wool and polystyrene.

Energy-efficient buildings in Poland typically include apartment blocks and offices. Solar energy can be used to shorten the heating season and supplement the heating systems. It is estimated that with appropriate design and renovation, energy demands for space heating can be reduced by 40%.

Current government policy focuses on the need to reduce domestic energy consumption. A two-stage programme is aiming to reduce net energy usage: in the early 1990s new houses will be required to achieve a net energy use of 120-160 kWh/m^2, and by the year 2000 this target will be lowered to 80-120 kWh/m^2. The success of the programme depends largely upon the development of a manufacturing industry for insulation materials. Building regulations and standards for insulation are also currently being amended and improved upon.

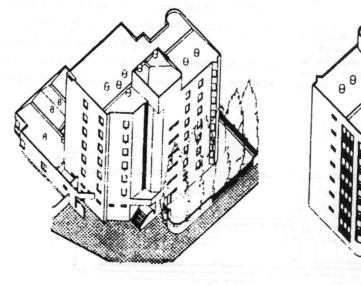

Above

Indicative graphic showing axonometric views of the north (left) and south (right) side of a 1988 design for an energy-efficient apartment block in Warsaw. The design, by Anna Czechowicz, Andrzej Janeczek and architects Jan Gorgul and Teodozja Neumann, was undertaken for a competition organized by the Technical Institute of Building Technology and the Association of Polish Architects, and gained an honorable mention.

Energy-efficient features include compact massing of the residential part of the building, glazed loggias used in conjunction with darkly coloured facades, and insulated skylights.

Although the development of hydro-electric power has enabled Turkey to satisfy its demands for electricity, energy supplies for space heating depend largely upon imported petroleum sources. There is a need to minimize the use of imported fuel types and a need to conserve energy by the use of renewable energy sources. Energy-efficient design is therefore becoming increasingly important.

Climate is varied, including hot dry, hot humid and temperate zones. Although winter heating is required in most regions and summer cooling is frequently also necessary, each geographic and climatic region has specific heating and cooling requirements. This necessitates a different set of energy-efficient design solutions for each region.

Building regulations intended to minimize energy consumption are under revision, although these are unlikely to meet the needs of all residential developments in Turkey, particularly the unplanned settlements which have emerged in recent years.

Turkey offers great potential for energy-efficient

designs which exploit site and orientation to maximize passive heating and cooling without resorting to the use of complex or costly technology.

A prominent example of an energy-efficient building following this approach is a primary school in Istanbul. Careful design for site, orientation and the positioning of glazed areas leads to maximum exploitation of solar gains during the heating season. The use of a courtyard linking the buildings together provides an open shaded area during the summer.

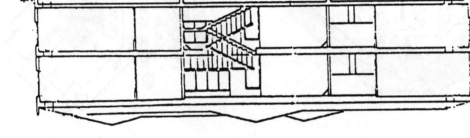

Above

Indicative graphic showing a section through an energy-efficient primary school in Istanbul. The school is divided into two groups of classrooms, each containing four classrooms per floor arranged around a central staircase. The two groups are linked by a central courtyard.

Energy-efficient features include careful orientation of classrooms and the use of a courtyard to provide shade and cooling. The school was designed in 1987 and completed in 1988.

yugoslavia

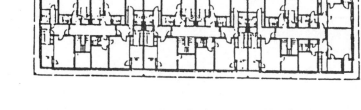

The lack of indigenous energy and a favourable geographical position make Yugoslavia a natural location for solar-based energy-efficient design initiatives. By the end of 1986, 18,000 solar energy systems had been installed, including 270 large plants owned by public enterprises with a total of 60,000m² of flat plate collectors, and 17,000 smaller installations with a total of more than 90,000m² of installed capacity.

The Yugoslav manufacturing industry has developed its own solar cells and is capable of producing them in capacities up to 10 kW. Wider application of these active solar collectors is at present hindered by high production costs, making them uneconomic for many building types and building users.

Solar heating and heat pumps are mostly used in the Adriatic coastal and south-eastern regions for the supply of hot water for dwellings, hotels, leisure and swimming pool complexes. Research shows that solar energy could meet 70-80% of the domestic hot water requirements in the coastal area and 50-60% of requirements in inland areas. Over 500 dwelling houses, all built since 1982,

incorporate passive or active solar design features.

Notable research and development is being undertaken in the computer modeling of building thermal performance to examine the thermal performance, heating costs, thermal balance, heat flows, air flows and temperature dynamics. One commercially-available programme is part of a computer-aided design package which combines 2- and 3-dimensional architectural modeling with evaluations of heating and cooling systems and computation of thermal loads.

Above
Indicative graphic showing plan, section and elevation of an energy-efficient apartment block, designed in 1988.

Yugoslavia has a far greater potential for energy-efficient design than has yet materialized. Many initiatives have foundered because of insufficient knowledge on the part of decision-makers, builders and designers about the technical and economic aspects of energy efficiency. Design and development activities are focused mainly on active water heating systems, heat recovery systems and heat pumps or computer modeling. Inadequate regulations at a national level means that legislation is the responsibility of individual republics. As a result there is a great variety of methods and approaches to energy-efficient design in different regions.

part four

computer software

section eleven   software for energy efficient design

software for energy efficient design

Energy-efficient design, although simple in conception, is in practice a sophisticated and complex process which involves many variable inputs and an infinite number of possible design solutions. Computer software helps designers simplify this process and test out the performance of their proposed designs.

Two types of software are available for energy-efficient design: those using simple inputs and outputs requiring less user expertise (such as calculators, spreadsheets or purpose-made micro-computer models), and those with advanced inputs and outputs requiring high levels of testing and interpretation (such as building simulations or expert systems running on workstations and mainframe computers).

The UNDP-ECE program which accompanies this guide is in the first of these categories, and may be run on virtually any computer supporting the MS-DOS operating system. Although primarily intended as a training aid, it is a proven practical tool which gives realistic calculations of space heating and total energy needs and associated costs. A two-zone variable-base

degree-day model accounts for heating and cooling demands, with the added facility of allowing national parameters such as climate to be changed as required. This helps designers to examine the fundamental variables that affect energy consumption in dwellings and to answer "What-if?" propositions about different designs. Data inputs are required for climate, site, dwelling type, building elements, volume of space, orientation, over-shadowing, sunspaces, and heating and hot water systems.

performance

software

design simplicity   design testing

Above
Three benefits of using computer software in the energy-efficient design process.

The UNDP-ECE program is an adapted version of the Milton Keynes Energy Cost Index commercially available as Energy Designer. The basic algorithms were developed and tested at the Building Research Establishment in the United Kingdom under the BREDEM programs series. Energy Designer is the copyright of Energy Advisory Services Ltd.

SOFTWARE FOR EED

UNDP-ECE software

This section gives a brief introduction to the UNDP-ECE energy-efficient design computer program and its origins, purpose and applicability.

Overview. The program evaluates the energy performance of a house design by calculating the amount and likely cost of energy used by the building. The program has facilities for diagnosing energy use, and evaluating the effects of changes made to the design. It is an adapted version of the Energy Designer program which has been in use in the United Kingdom since 1985. The program has been calibrated and tested for houses in the United Kingdom and performs reliably. Results cannot be guaranteed if the program is used for other building types.

Two features have been added to Energy Designer: firstly, a configuration module, which allows the program to be configured for use in different countries; and secondly, cooling calculations, which enable the program to be used in hot countries where cooling is necessary.

The program has been designed to run on any type of computer with the MS-DOS operating system, and to

be easy to use. The main part of the program requires input data on the areas and U-values of external building elements and data on heating systems and controls. The space heating load is calculated, together with all other fuel requirements, using standard assumptions about occupancy, heating requirements, water use and so on. The total fuel cost is calculated and a simple cost index is derived.

In all there are just over 50 items of input data required by the program, but once the data have been entered they can be saved and the program re-run with new variable settings through a revision facility without needing to re-enter the data. All data input requirements are prompted, and there is a Help facility that can be used at any time during input or any other program operation. If U-values are not known, the program will estimate them. Many items can be chosen from on-screen menus. The program incorporates a diagnosis facility to help with any breakdowns in the energy calculations.

The energy calculation model is a variable-base two-

zone degree-day model. The basic algorithm has been developed from extensive field research carried out at the UK Building Research Establishment, so the model has been calibrated against live performance data. The program incorporates routines for dealing with very low energy houses.

Sequence of operations. The program will automatically save all the data that you enter on the disk, but you must take precautions to make sure that your data are not over-written or deleted. First, create your own data disk; secondly, ensure that this data disk is regularly backed-up. This will require knowledge of the MS-DOS formatting and copying facilities, and a regular routine for disk management. You will need formatted disks for making a back-up copy of the program disk, storing the data and making back-up copies of the stored data. The FORMAT and COPY utilities of MS-DOS are briefly explained in the section on creating a program disk. For more detailed explanations consult your own MS-DOS manuals.

continued

Installation. You will need to install the program on the computer if this has not been already carried out. The program may be run from disk or installed on the hard disk, if you have one. If you are running the program from floppy disks the program will require two disk drives; one for the program and operating system (normally drive A) and the other for the data (drive B). The program disk does not contain the operating system which is needed to run it because this is a copyright program which is normally supplied with the computer. This means that you will have to create a working copy of the program by using a formatted disk onto which is copied the operating system and the program.

If you are using the program on a hard disk (which should already contain the operating system) you will need to install the program on the hard disk. Note that when the program is installed on a hard disk it can quickly fill the disk with data files, so it is usually better to write the data to a floppy drive.

Running the program. The next step is to run the program so that you obtain the introductory menu. The program will run by typing "UNDP" (without the inverted commas) in response to the MS-DOS dot prompt. You will need to make sure that you are using the drive on which the program disk is located, otherwise the program will not run. On a twin floppy disk computer, this will normally be drive A; on a hard disk it will normally be drive C. You may also run the program from a floppy disk on a computer with a hard disk. To switch to the disk drive you require type the disk letter, followed immediately by a full colon in response to the dot prompt (thus A: typed after the dot (.) prompt switches to drive A. The introductory menu is the first menu on the screen. It has five items:

1   Run Energy Designer
2   Introduction and help
3   View or change configuration
4   Hardware set up
5   Quit

Item 1, "Run Energy Designer", runs the main program. When the program is run for the first time, though, you should choose item 3, "View or change configuration". This enables users to configure the program for use in their own country. This configuration may have already been carried out for you, but if it has not you will need to run this item yourself and input the necessary data. The configuration supplies the program with climatic and cost information. This information differs from country to country, so it must be included for the program to run correctly for your own geographical location. Item 2, "Introduction and help', gives access to all the help information. This option may be run as an introduction to the program. Item 4, "Hardware setup", enables the hardware such as the printer to be set up for running with the program. This is covered in the section "Introductory menu: hardware setup". Item 5, "Quit", returns to the operating system.

Once the configuration has been carried out, Energy Designer may be run (Item 1 of the Introductory menu). You will be asked to provide a job title, an optional comment, a version number (1-99), the date and a filename. The filename will include the version number so the same filename can be used for different versions of the same job. The program will then ask for site data and general information about the house. This is used for the calculation of U-values and ventilation rates.

At the end of the input process the program calculates the amount and cost of energy used by the dwelling. Once the basic data for the dwelling have been typed in, they are saved to disk and can be used again. The program has diagnostic features to examine the calculations and allows "what-if?" questions to be asked of alternative design solutions.

menu structure

The menu structure of the UNDP-ECE version of the Energy Designer computer software is arranged as shown opposite.

From the main menu, you have the option of running the Energy Designer program, getting introduction or help, viewing or changing the configuration, establishing hardware characteristics, or returning to the operating system. The following pages explain the purpose of each of these options, the thinking which lies behind them, and what you will need to do in order to make the program run correctly. You should pay particular attention to the instructions about configuration as any mistakes made here will impair the appropriateness of results for your own country or region.

In addition to the menu options, we provide the fundamental algorithms used by the program to perform its energy efficiency calculations.

Energy Designer

version 2.10
October 1990

introduction and help

1 introduction to system
2 help system
3 Energy Designer
4 hardware set up
5 configuration files (general)
6 quit (to main menu)

main menu

1 run Energy Designer
2 introduction and help
3 view or change configuration
4 hardware set up
5 quit (exit program)

view or change configuration

1 create new configuration file
2 recall existing configuration file from disk
3 directory of existing files
4 view configuration parameters
5 revise configuration data
6 select file for use by Designer
7 save new configuration file to disk
8 print current configuration file
9 help and information on configuration
10 calculate degree days
11 quit (to main menu)

hardware set up

1 monitor type
2 print control codes
3 data location
4 save new file
5 quit (to main menu)

main menu: Energy Designer

Energy Designer is the main part of the UNDP-ECE program. The program requires a detailed set of data about the dwelling. When the program is first run all the data have to be input manually. It is essential to have a plan of the dwelling available with dimensions marked on it.

Input sequence.
This section provides details on the sequence of input data.

- **Set help level**. The help level determines how much information is provided for the user. Choose level 1 if you have not used the program before, otherwise use level 2.

- **Recall job from disc**. This allows you to use a previously saved set of data. Answer N for a newly created job. The files PENNYLND.V1 and PENNYLND.V2 are included as examples of UK buildings. To recall these files answer Y to the prompt and type "PENNYLND".

The system will then prompt you to give the version number, and provide a new job name. You will then be taken into a revision procedure so that you can change previously entered variables, if you wish.

Pennyland is the UK reference building for energy efficient design. It provides a baseline for all other designs.

- **File names**. You will be asked to provide filenames, dates and version numbers to identify your jobs. A filename must not have more than 8 characters. Use letters and numbers entered with the CAPS LOCK key on (for example, "UNTEST12").

- **Site data**. Type in the soil type - gravel, clay or other.

- **Wind shielding**. Ventilation rates are affected by the extent to which the building is surrounded by external structures. Type in the wind shielding - no obstacles within the distance of twice the height of the building, few obstructions, obstructions on 2 or 3 sides, obstructions on all sides. Type in whether there are any obstructions to the south-west.

- **House information**. This is the type of dwelling: detached house, semi-detached house, end terrace house, mid-terrace house, end terrace in back-to-back arrangement, mid-terrace in back-to-back arrangement, and flats (apartments). Note that the program applies to single flats, not to apartment blocks as a whole.

House use covers whether the dwelling is a normal family dwelling or a home for old people. Bed spaces is the planned number of occupants.

- **Building elements**. This section requires input data for the heat loss elements of the building. The program will accept up to two different types of construction for each of external walls, roofs and external floors (floors with heat loss areas on the external face of the building). Elements with different U-values

continued

should be entered separately. Elements with different construction but the same U-value can be entered as one type. If you have more than two types, the most similar elements should be combined and the averaged value used.

• **Zones.** The two-zone model requires that zone 1 is the living area, making the living room and dining room effectively one room. The rest of the dwelling is zone 2. The kitchen can be included in either zone 1 or zone 2, depending on use and location.

• **U-value data.** You may either type in a U-value for a building element, or the program will calculate it. U-values are in Watts per square metre. The program will request the U-values for wall types, referring back to the wall types entered previously.

• **Area data.** All areas are in metres entered as a series of length by breadth statements. Always enter the gross areas. Do not subtract the areas of the opening such as doors and windows. This is done automatically later in the program. Enter area data by typing first the length of an element, then press <ENTER>, then the breadth the element. You may enter a series of dimensions in a list on the screen. At the end of the sequence press <ENTER> twice. The total area for the element will be calculated by the program. If you make a mistake type R to revise and you will be able to edit the entry.

• **Ground floor U-value.** The program requires the thickness of the ground floor insulation. Type 0 if there is none.

• **Openings.** You are asked to enter the main orientation of the glazing of zones 1 and 2. Choose "other orientations" unless the dwelling is on a north-south axis (within 30 degrees). With such dwellings choose the option on the basis of the proportion of the glazing on the south face.

In order to estimate the overshadowing of a glazed area imagine standing one metre away from the glazing inside the building. Estimate how much of the visual field of the window appears as sky and use this to choose the overshadowing.

Opening areas are entered as pairs of dimensions like walls. You are also asked for a code for the type of wall or roof in which the opening is located, the type of opening and a code for draughtiness. Typing "S" repeats the entry of the last opening entered.

• **Double glazing.** If some of the openings are double-glazed the program requests their details. If they have special transmittance due to coating you must enter the details.

• **Heating systems.** Where there is one heating system this is the primary system. If there are two systems make the primary system the one that provides heat in both zones and hot water. If the heating system is not shown on the menu choose "Other". Responsiveness is the speed with which the system increases or decreases its heat output in response to consumer or automatic control.

• **Miscellaneous input.** Low energy lights and appliances, gas cooker point and location of the kitchen affect incidental gains. The draught lobby on the main door (an enclosed porch) reduces the air change rate. The mechanical ventilation system replaces the occupant's openings of windows by a mechanical air change system with heat recovery on the exhaust air. A loft hatch (a ceiling access door or panel) affects the stack-effect air flow. In the number of chimneys do not count balanced flues (boiler flues which draw in as much air as they exhaust).

If you answer yes to the sunspace/conservatory question you will be asked to input data on the details of house elements covered by the conservatory, the orientation, the fraction of the total house area covered by the conservatory, the gross dimensions of the conservatory elements, inclination, orientation, overshadowing and element type.

Output sequence.

The program calculates an energy cost index, the total delivered energy and the total primary energy. A fuel

continued

analysis breaks down the fuel consumption according to application (space heating, cooking, lights and appliances and water heating) and by fuel type. A routes menu then provides the possibility of saving, revising, diagnosing, printing, changing the version number, resetting the help level and of running another job.

- **Energy cost index (ECI)**. The index is calculated from the estimated total fuel costs, including standing charges and total maintenance costs, per unit floor area. The index is constructed in terms of the extra fuel running costs over a minimum requirement which is set at approximately the annual fuel bill for a very small, one-person all-electric flat.

Fuel use is multiplied by current prices, with maintenance charges added. This provides a total fuel cost. This total cost is adjusted by a function which takes into account the price of fuels. This process ensures that the ECI of a standard house is always 100, no matter the actual fuel price. This is the adjusted total cost. A fixed minimum energy cost sum (in the UK it is £150) is subtracted and the result is then divided by the floor area and multiplied by a fixed scaling factor to obtain the index.

- **Fabric index**. The fabric index is the space heating requirement as designed, divided by the space heating if built to building regulations multiplied by 100. It is assumed by the program that the building regulations

version of the house has the following settings:

Wall U-value	0.6
Roof U-value	0.35
Glazing U-value	5.0
Floor U-value	Calculated for house shape with no insulation installed
Air change rate	Calculated for house with no draughtproofing or draught lobby.

The space heating requirement is the net useful energy demand, that is, the useful heat demand minus the expected contribution from internal and solar gains. It does not include any appliance or heating system efficiency. The index is seen in the UK as a valuable device for marketing good energy performance.

- **Total delivered energy and primary energy**. These are shown in gigajoules (GJ). To change energy units to kilowatt hours type K.

- **Fuel use analysis**. This is provided by application and fuel type. If the results appear to be incorrect the diagnosis routine may be run to check their validity.

- **Diagnosis**. Diagnosis (item 3 of the routes menu) allows analysis of total fuel use, heating demand and efficiency, hot water system, incidental gains, specific loss and U-values. Use diagnosis for all dwellings entered.

main menu: introduction and help

Introduction and help is item 2 of the main menu. This section is an overview of information contained in the help environment. The UNDP-ECE system provides guidance on the energy-efficient design of dwellings by evaluating the space heating and cooling energy requirements. Once a design has been entered into the computer-program, a large number of variations can be tried and evaluated. As well as space heating and cooling energy, the program also estimates the energy consumed by cooking, water heating, lighting and appliances.

Energy Designer. This program, first produced in 1984, has been used to evaluate and improve the design of more than 4000 energy-efficient house and apartment block designs in the United Kingdom. The core algorithm is based on BREDEM - the Building Research Establishment Domestic Energy Model. This is a two-zone, variable-base, degree-day model developed and verified using data from dwellings in the U.K.

Although the program can be configured for non-domestic dwellings the core algorithms have not been verified for these building types. The UNDP-ECE version

of the program has a simple degree day cooling algorithm included for estimating cooling loads.

The system must be configured for each country and region in which it is used. To configure the system, data must be entered via the configuration option on external temperatures, solar flux, hot water, fuel costs and other items. Once these are entered, the configuration file can be saved and used in future sessions.

Using help. This screen is part of a general help facility that can be accessed throughout this program. There are two methods for getting help. The first is to select the help option on those menus where it is an option. The second method is to press the F1 key at any point where a menu choice or some data entry or some other keyboard response is required. Each section of the program has its own help system and help screen - so when you press F1 you will get information related to that part of the program. This also means that if you want further information on some part of the program menu - and press F1 to see the appropriate help screen.

main menu: configuration

The view or change configuration is item 3 of the main menu and sets all of the program parameters which vary from country to country and from region to region. When the Energy Designer program starts it reads in this data so that the results of the calculations are correct for that particular region.

Data are entered using the options in the configuration section of the Energy Designer program. Once the data have been entered they are then saved to disk.

The user can create as many configuration files as are required. However, only one of these files will be used in the Energy Designer program during an evaluation. Files which have been previously created can be recalled, edited or examined, and selected for use by Energy Designer.

The file to be used is copied into the file DESIGNER.CFF which becomes the file subsequently read by the Energy Designer program. The copy operation is done automatically. You can also use the normal MS-DOS COPY command to accomplish the same result.

Note. The data files created and stored by Energy Designer do NOT contain configuration data; they contain only dwelling data. The same dwelling data file can lead to different results if evaluated using different configuration files. This makes it possible to evaluate the same dwelling in different climatic regions. But it also makes it easier to forget how a particular set of results were obtained.

The **Create** option enables you to produce a new configuration file. You are taken through a sequence of data entry routines to specify all the data required. Data must be entered for the following.

- **Temperature data and degree days.** The program calculates the degree days on the basis of the average external temperatures for each month of the year. You should enter the temperatures, in degrees Centigrade, for each month starting with January and ending with December.

The degree day calculation also requires one further parameter, denoted k. The value of k is $\sqrt{(2\pi)}$ divided by the standard deviation of the external temperature about the mean value.

If you do not know the variation of the temperature but have information on the degree days for a given base temperature then you can obtain an approximate value for k by trial and error. The option "Calculate Degree Days" on the Configuration Menu provides this facility and enables you to adjust k to get agreement with the known degree days.

- **Solar flux**. The solar flux values, in Watts per sq. metre, are required for the estimation of solar gains. The program has its own routines for correcting for different transmission factors for different degrees of overshadowing and for the ratio of glazed area to total window or door area. The values to be entered are therefore the raw flux values not corrected for transmission through windows in any way. Two sets of values are required. Each set requires flux values for vertical surfaces facing south, north, and east or west (taken to be the same). The first set requires the flux values to be averaged over the heating season; the second set averaged over the cooling season. The flux values also have to be averaged over 24 hours. The heating and

continued

cooling seasons used for the averaging should be the same as those entered with the temperature data.

• **Cooking**. The program requires accurate assessment of all auxiliary fuel use in order to estimate correctly the internal gains. In low-energy dwellings, gains may easily contribute more than 30% of the total heating requirement, and may make a similar contribution to the need for cooling. The use of fuel in cooking is estimated in two stages. The first stage estimates the fuel used in an electric cooker using the equation

cooking = A + (B x nump) GJ/yr

where A and B are coefficients and nump is the number of occupants. The configuration program requests values for A and B. The actual values for the U.K. are given in below in the section on calculation algorithms.

Note. Setting B equal to zero means that the cooking fuel is independent of the number of occupants. For most countries there is a relationship between the cooking fuel used and the number of occupants. For the U.K. B = 0.5 for cooking by electricity, a figure which has been derived empirically.

The second stage of estimating corrects for the relative efficiency of different cookers. The program assumes that gas cookers require 1.5 times as much fuel and coal, and oil-fired cookers twice as much fuel, as electric cookers.

• **Electricity use**. Electricity use has to be estimated correctly to obtain a reliable estimate of internal gains.

Electricity consumption by lights and appliances can be predicted in three ways. The first way is to base the estimate on the number of dwelling occupants (nump). The second way is to base the estimate on the total floor area (tfa). The third way is to use their product (nump x tfa). Select the method suited to your data to estimate the corresponding coefficients. The equation used to estimate electricity includes terms for each approach - usually only two coefficients are not zero.

Note. Electricity used in space and water heating and in cooling should NOT be included in this category - the estimate required here is related solely to electricity used in lighting and general appliances.

• **Water heating**. The water heating model used in Energy Designer needs an estimate of the energy required to heat the volume of water used as hot water from the cold input temperature to the temperature at which the water is used or stored. Volume is estimated by

volume = a + b x nump cu.m/yr

where a and b are constants and nump is the number of occupants. This can be converted to an energy requirement by multiplying water density (998 kg/cu.m), its specific heat capacity (4190 J/kg) and the required temperature rise (T[hot] - T[cold]). This gives the equation

energy = A + B x nump GJ/yr

The program requires you to enter the value of the coefficients A and B.

The program accounts for solar panels in the water heating model. It cannot deal with panels used for space heating. Solar panels are assumed to be linked to the water heating system thereby providing a fraction of the total heat required - usually in a pre-heat function. Useful heat output from the solar panel is dependent upon its area (Area) and upon the demand for hot water (HWD - estimated separately). The greater the area the greater the output. In addition, the greater the ratio of demand to output the greater the useful output. Output is taken as

output = k x Area

and the overall useful output is taken as

useful output = k x Area (1 - f(HWD/k x Area)) GJ/yr

where f(HWD/k x Area) is a usefulness function.

The function used in the program is:

1/(a + b x HWD/(k x Area))

The coefficients k, a and b have to be specified.

continued

- **Fuel costs**. Fuel costs are required in order to calculate the main performance measures used for assessing the dwelling energy efficiency. The program requires a symbol for the currency unit to be used (this may be up to three characters long). It also requires the fuel costs in the local currency unit per GJ of energy contained. In order to convert from costs per unit of fuel to costs per GJ the cost per unit must be multiplied by the number of units per GJ. For example, if electricity costs 0.10 $/kWh then, since there are 277.8 kWh per GJ, the cost per GJ is

277.8 x 0.1 = 27.78 $/GJ.

For oil, gaseous and solid fuels the conversion from natural units to GJ depends on the fuel calorific value.

- **Occupancy conditions**. The system requires a standard set of heating and cooling occupancy patterns to evaluate the energy requirements. It is assumed that the whole dwelling is heated (or cooled) for one or two periods per day. If there is only one period, enter zero for the length of the off period and for the second period in the input routine. The demand temperature for heating (or cooling) is the temperature that the heating (or cooling) system aims to achieve during the occupancy period(s).

The program also includes the following configuration options:

- **View** provides a summary of the data in the current configuration file.

- **Revise** allows you to change any of the data in the current configuration file.

- **Print** allows you to obtain a hard copy print out of the current configuration data.

Note. Before you can view, load, review or print you must create a configuration file or load one from disk (using the Recall option.

- **Save** saves the current file to disk.

- **Calculating Degree Days**. This function has been provided so that the operator can check that the entered temperatures and degree day coefficient leads to an accurate estimate of the degree days.

In order to carry out the check you need to know the degree days for your region for a specified base temperature, and for a known heating season (number of months in the year). Set the heating season in the program to be the same as the heating season for the known degree day value (hence for an annual degree day figure set all months to heating). Select "Calculated degree days" and enter the base temperature which was used for the known degree day figure. The calculated value should then agree with the known degree days. If it does not agree, check the temperatures or adjust the value of k. (For the U.K., for example, k is 0.71).

The degree days for heating and cooling are calculated using the external temperatures provided and the coefficient related to the standard deviation of the temperature distribution. This method works accurately if the length of the heating and cooling seasons are correctly specified. Without specifying the heating season length, the calculated degree days include small summer contribution, even though heating systems are not used in this period. Months when no heating is used should therefore be excluded from the heating degree day calculation. The same applies to the cooling degree days: exclude those months when cooling systems are not used.

SOFTWARE FOR EED

main menu: hardware set up

Hardware set up is item 4 of the main menu. It allows you to establish the following.

• **Monitor type.** The program can operate in monochrome or colour. This option allows you to specify which is appropriate for your computer system. The program can automatically sense whether the video adapter has a colour capacity or not. However, many system make use of various shades of brightness to mimic a true colour monitor. For such systems some of the screens in the program may be difficult to see, so it is best to select the Monochrome option for these systems. It is also best to select the Monochrome option for portable computers with LCD or gas plasma displays.

• **Printer codes.** The program comes configured for an Epson FX80 series printer. Many non-Epson dot-matrix printers can be set to emulate this machine. If you wish to enter the specific codes for your printer then you will need to consult the printer manual to find the codes required to start and stop enlarged letters, start and stop bold (or emphasized) printing, the codes for form-feed and any initialization codes that should be sent to the printer. You require the ASCII decimal equivalent of each of the code symbols (for example, the Esc code required is 27). If you have difficulty getting an adequate print out then start by setting all the codes except the form-feed code to zero. This produces a simple, legible print out.

• **Data location.** This specifies the location where Energy Designer will save its data files on dwellings. This can be specified as any valid drive or directory on your computer system. If you specify a sub-directory then be sure to create it before using the Energy designer program, otherwise data will not be saved. If you specify a floppy disk drive as the data location then ensure that there is a formatted floppy disk in that drive with sufficient space to store the data files.

Each section in the hardware setup has instructions which are presented to you before you enter any new values or data. The values which you enter are stored in the file DESINIT.HFG which is read by both this program and the Energy Designer program.

• **Save new file.** Whenever a change is made to the hardware set-up it will be used for that duration of that session using the program. If you wish the new set-up to be used next time the program is run then you must save the new hardware set-up to disk.

System requirements.

Energy Designer will run on an IBM PC with or without a hard disk. However, for convenience of use, the programs are arranged differently on the hard disk machine.

Normally, the Energy Designer program requires two disk drives, one of which is used to hold the program, and the other is a data storage disk. This is the way the system is set up on the twin-disk drive IBM PC. The program does not require any special configuration.

For the hard disk version (IBM XT) the arrangement is different since the programs are stored on the hard disk

continued

leaving the floppy disk for storage. The following applies mostly to the twin floppy versions of the programs. For installation on hard disk machines the instructions are extremely simple since operations are carried out by batch file.

Note. With the hard disk version it is possible to arrange for the program to save data on the hard disk. In general, this is not good practice since the hard disk will quickly become full of data files. However, for dedicated machines this is satisfactory and you can make the necessary changes using the install program.

Creating a program disk.

If you have a hard disk version of the program then you should switch on as normal and get to the C> prompt. Then place the Master disk in drive A and type A:HARDDISC. A batch file automatically creates a sub-directory, transfers the program files to the directory and then places a batch file in the root directory which enables you to start the program by typing UNDP.

If you are installing the program on a twin floppy machine then the procedure is more complicated. The disks you receive are not operating disks since they do not contain copies of the MS-DOS operating system nor the RAM-disk program recommended for use with the program; these are both copyright programs and usually come with the computer. Thus to create a working disk you will need to add copies of your version of the operating system. You have to do this on new disks - keep the supplied disk as a Master disk and always work with copies made from it. The following instructions tell you how to do this.

The first step is to format some blank disks. This is done using the format utility.

BEWARE. The format utility wipes ALL data from disks. Do not use it while master disks are in either disk drive.

You will need at least one storage disk and one program disk. The program disk will need the system tracks copied onto it; this is not necessary for storage disks. In order to format a disk ready for use as a storage disk place a DOS Utilities disk in drive A and a blank disk in drive B. Switch the computer on if not on already. Then type

FORMAT B:/S <RETURN>

To format a storage disk place a fresh blank disk in in drive B and type

FORMAT B: <RETURN>

Note. If you wish to place "volume labels" on your disks then you must add /V to both of the above commends and enter the label when prompted to do so.

Once you have finished formatting as many disks as you require then exit the format utility by responding N(o) to the question "Do you wish to format another disk Y/N"

Now place a newly formatted disk that contains the system tracks (formatted with the /S option) in drive B and place the Master disk in drive A. Then type

COPY *.* B: <RETURN>

Both disk drives will activate and you will see a list of the files as they are copied from one disk to another. The files copied in this operation are as follows:

INSTALL.DAT	cost data etc
AUTOEXEC.BAT	start up file
UNDP.EXE	configuration program
WORDS7.TXT	text file for menus etc
HELP7.TXT	text file for help messages
INSTALL.EXE	alters set up and costs
LOADER.EXE	loading program
UHELP.TXT	
BRUNIO.EXE	
DESINIT.HD	
UKNORMI.CFG	U.K. configuration data
DESIGNER.EXE	Energy Designer source code
DATA <DIR>	data files

When it has finished you should see the message "n file(s) copied". If there are any error messages then repeat the operation until it is successfully completed. Repeat this operation for each program disk you wish to create.

calculation algorithms

The following sections set out the most important equations and algorithms used by the UNDP-ECE version of Energy Designer.

- **Heat-load calculation.** The basic equation for the heat transfer between a house and its surroundings can be written as

$$q + G = (\Sigma AU + Cv)(T_i - T_o) + S$$

where

q = rate of heat output from the heating system

G = rate of incidental gains

ΣAU = fabric specific loss per degree

Cv = ventilation loss per degree

T_i = internal temperature

T_o = external temperature

S = rate of heat transfer in or out of storage in house structure

All terms in this equation vary continually over time. The equation can be integrated and expressed in terms of 24 hour mean values.

$$\bar{q} + \bar{G} + (\Sigma AU + Cv)(\bar{T_i} - \bar{T_o})$$

The heat storage term will normally be negligible and has been dropped from the equation on the assumption that thermal capacity effects will cancel out when the heat requirements for each day are totalled over the heating season. If we denote the total specific loss $\bar{U}$ by

$$\bar{U} = \Sigma AU = Cv$$

then we get

$$\bar{q} = \bar{U}[\bar{T_i} - (\bar{G}/\bar{U}) - \bar{T_o}]$$

To obtain the annual space heating requirement this equation is integrated over the heating season. A good approximation to this integral will be the summation of the term $[\bar{T_i} - (\bar{G}/\bar{U}) - \bar{T_o}]$ for days on which

$$\bar{T_i} - \bar{G}/\bar{U} > \bar{T_o}$$

This is close to the conventional definition of degree days to the base $\bar{T_i} - \bar{G}/\bar{U}$. Denoting this by the function DD( ) we have the result as

$$Q = 8.64 \times 10^{-5} \, \bar{U} \, DD(\bar{T_i} - \bar{G}/\bar{U})$$

The constant 8.64×10^{-5} is necessary to express the final answer in gigajoules (GJ). The necessary degree day tables have been incorporated into the program. This provides the variable-base degree-day part of the model.

The other part of the model is the two zone model (see illustration below). Each zone has its own specific loss U_1 and U_2, and there is a heat transfer between the two zones governed by the specific loss term U_3. Each zone has its own heat inputs as incidental gains, G_1 and G_2, and auxiliary heat input from heating devices, q_1 and

continued

q_2. The temperature of each zone may be different and is denoted by T_1 and T_2 and the external temperature is T_o.

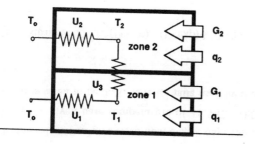

This system of heat flows is solved by assuming some knowledge of the temperature in zone 2. Even in houses where there is no explicit temperature controller in zone 2 it can be reasonably assumed that the occupants will make adjustments (by balancing radiators or by turning them on or off) so as to obtain a zone 2 temperature close to the design temperature.

Any overheating caused by the lack of automatic control is estimated as a function of the heat loss parameter (HLP - the specific loss per unit floor area). Essentially, the overheating is assumed to be negligible for a HLP value of 6. For an HLP value of 0 the temperature in zone 2 is assumed to be the same as that in zone 1. When temperature controls are introduced into zone 2 the magnitude of the overheating term is halved.

A knowledge of the interzone heat transfer

coefficient U_3 is also required. This can be calculated by the equation

$$U_3 = 23 + 3.18F \text{ TFA}$$

where

TFA = total floor area
F = fraction of total floor area occupied by zone 1

Where the temperature in zone 2 is uncontrolled, heat load is given by the equation

$$Q + 8.64 \times 10^{-5} \frac{(1 + a)}{(1 + aU_D)} (U_1 + U_2U_D) \text{ DD} \{T_1 - \frac{(G_1 - U_DG_2)}{(U_1 - U_2U_D)}\}$$

where

$$U_D = U_3/(U_2 = U_3)$$
a = ratio of heat output of zone 2 to zone 1

Where the temperature in zone 2 is controlled, heat load is given by the equation

$$Q + 8.64 \times 10^{-5} (U_1 + U_2U_D) \text{ DD} \{T_1 - \frac{(G_1 - U_DG_2)}{(U_1 - U_2U_D)}\} +$$

$$8.64 \times 10^{-5} (U_2/(U_2 + U_3)) \text{ DD} \{T_2 - (T_1-T_2) U_3/U_2 - G_2/U_1\}$$

The evaluation of the solutions given above requires knowledge of the following eight terms.

T_1 = mean internal temperature in zone 1
T_2 = mean internal temperature in zone 2
U_1 = specific heat loss coefficient
U_2 = specific heat loss coefficient
U_3 = specific heat loss coefficient
a = ratio of heat into zone 2/zone 1
G_1 = incidental gains in zone 1
G_2 = incidental gains in zone 2

• **Mean internal temperatures**. The mean internal temperature can be visualized as the average of the temperature time graph for a particular zone. The graph below is for a zone heated twice a day.

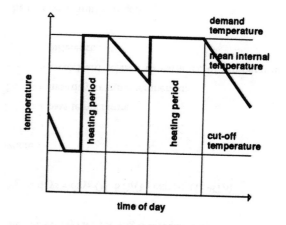

The graph is specified by specifying a demand temperature, a cut-off temperature, the number of hours

continued

of heating and the hours between periods. Given this data and the assumption that the variation can be reasonably represented by the type of graph shown, then the average can be calculated by using the methods of geometry.

- **Degree days**. In order to avoid over-estimation of heat requirements for houses heated with unresponsive systems, and in order to cope with the practice by which people tend to turn off their heating systems in the summer months, the program excludes four summer months (June, July, August and September) from the degree day calculation.

In effect this means using 8-monthly degree day figures instead of annual figures. The degree day total is scaled in proportion to the degree day base for the country or region in question. The external temperatures used in the calculation give an annual degree days value of 2500 for a given base temperature of 15.5°C. The coefficient is taken to be 0.71. The formula used by the program is

DDays = (ddbase/2500)* N*(T^b - T^m) / (1 - exp)-k(T^b - T^m))

where

T^b	= base temperature
T^m	= average external temperature
N	= number of days in the month

- **Cooling load**. The degree days for cooling are calculated with a given "demand" temperature above which cooling is required. One strategy of cooling for sealed building in which cooling systems operate whenever the external temperature is near or above this demand temperature. The other strategy is for naturally-ventilated buildings in which ventilation rates are increased to reduce overheating, or a cooling system is switched on.

Under cooling conditions the mean internal temperature is calculated assuming that the house is cooled during the demand periods and the temperature is allowed to rise outside those periods. Thus the mean internal temperature will be greater than the cooling temperature. The house will require a cooling input when the average external temperature exceeds the desired mean internal temperature. Since there will be a temperature rise caused by internal and solar gains, cooling is actually required when the external temperature rises above the mean internal temperature minus the temperature rise caused by the gains.

The temperature rise caused by the gains is the ratio of the gains divided by the specific loss. Note that the specific loss includes the ventilation loss and that this will vary for the different cooling strategies described above. The dominant gains will be solar gains. Note here that the over-shadowing in the summer may be quite different from the winter over-shadowing due to the seasonal angle of the sun. The cooling calculation is

cooling load = constant x spec. loss x degree days cooling

The degree days for cooling is calculated by

degree days = days (TE-BT)/(1 - exp (-k (TE-BT)))

where

BT	= base temperature
TE	= monthly external temperature
k	= a constant related to variance of temperature difference

The precise definition of k is

k = $(2\pi/\sigma)^{0.5}$

where

σ	= standard deviation of external temperature about its mean value

The program operator must supply a value of k - this can be calculated from the known variance of the external temperature.

- **Specific loss**. The specific loss for a building zone equals ventilation loss plus fabric loss. The fabric loss for

continued

each zone is equal to the sum of the area of each element multiplied by its u-value. Ventilation loss is described in the next section. This section deals with fabric loss.

Areas for all building elements are entered by the operator. For walls, roofs and non-ground floors u-values may be entered directly or calculated. For ground floors u-values are always calculated. For windows and doors, standard u-values are built into the model. U-values for up to two materials up to six layers can be calculated by

$$U = \frac{1}{0.18 + \Sigma R(i)}$$

where

U $\quad$ = u-value
R(i) $\quad$ = resistance of the ith layer
0.18 $\quad$ = surface resistance

The resistance of each layer is calculated by

$$R(i) = \frac{T}{C_1(i) + f *(C_2(i) - C_1(i))}$$

where

T $\quad$ = thickness of the ith layer
$C_1(i)$ $\quad$ = conductance of material 1
$C^2(i)$ $\quad$ = conductance of material 2
f $\quad$ = fraction of cross-sectional area occupied by material 2

For air gaps greater than 27 mm a fixed resistance is used (equal to $0.18m^2{}^o CW^{-1}$ for normal gaps and $0.90m^2{}^o CW^{-1}$ for low emissivity gaps). For air gaps smaller than 27 mm, the equation used is

$$R = 1/[4.63 + (25/t)]$$

where

t $\quad$ = thickness of the air gap

Ground floor values are calculated by

$$U = 1/(R + (25 * t))$$

where

t $\quad$ = thickness of insulation below the floor slab
R $\quad$ = uninsulated resistance

The uninsulated resistance R is calculated by

$$R = \frac{\pi B}{4\,E\,exp\{B/2L\}\,arctanh\{B/(b + W)\}}$$

where

B $\quad$ = smallest slab dimension
L $\quad$ = slab length

W $\quad$ = wall width (taken as a constant 0.3 m)
E $\quad$ = earth conductivity (gravel 0.7, clay 2.1 and other 1.4)

• **Ventilation loss.** The basic model equation can be expressed as

$$ach = shelt * (i + v)$$

where

ach $\quad$ = air change rate
shelt $\quad$ = shelter factor
v $\quad$ = ventilation
i $\quad$ = infiltration

The shelter factor gives on overall reduction of 30% from a fully exposed to a fully sheltered site position. Infiltration is evaluated in terms of the building components. Ventilation arises from deliberate opening of doors and windows, and is calculated in order to maintain an overall total air change rate. The basic relationship is calculated in the algorithm as

$$v = 0.6 - 0.6 * (i)$$

subject to the constraint that it should not fall below zero. With mechanical ventilation the air change rate,

continued

93

taking into consideration heat recovery, is calculated as

$$ach = shelt * (i) + 0.17$$

• **Incidental gains.** To accommodate the degree to which incidental gains are useful, the "usefulness" of internal and external solar gains is estimated using the ratio of the total gains entering the house divided by the specific loss. The equation used is

$$usefulness = 1 - exp(-18 \text{ x spec loss/gains})$$

Metabolic gains are taken as 1.5 kWh/day per person. The number of people is calculated from the total floor area, and the gains are evenly divided between zone 1 and zone 2.

Cooking gains are taken as 90% of the fuel used for electric cookers and 65% of that for gas cookers. The fuel used depends on the number of people. For gas cooking the equation for the U.K. is

$$cooking\ fuel = 3.5 + 0.7 * nump \text{ GJ/yr}$$

and for electric cooking it is

$$cooking\ fuel = 3.2 + 0.5 * nump \text{ GJ/yr}$$

Cooking gains are allocated to zone 2 unless the kitchen is specified as being in zone 1.

Hot water gains originating from losses from the pipes between the boiler and tank are taken as 5% for normal systems and 15% for thermosyphon systems. Hot water gains originating from tank losses are calculated from the thickness of tank insulation. Hot water gains originating from conduction and convection losses from hot water used is taken as equal to 10% of the energy used in hot water. The energy used in hot water is taken as

$$1.27N + 2.02 \text{ kWh/day}$$

where

N   = number of occupants

Any solar panels are subtracted directly from this demand.

The consumption level of lights and appliances is estimated on the basis of total house floor area (TFA). The equations used are

$$electric\ gains = 2.68 * TFA \text{ W}$$
$$electricity = 0.064 * TFA \text{ kWh/day}$$

The total electricity is split 60% to appliances and 40% to lights. The gains are divided evenly between zone 1 and zone 2. For appliances only 0.7 kWh/day (a television) is normally assumed for zone 1, the rest being in zone 2.

However, if zone 1 includes the kitchen then about 80% of appliance gains (plus cooking gains and 50% of hot water gains) are allocated to zone 1. For an off-peak system, 10% of appliance electricity use is assumed to be consumed at the off peak rate. Where low energy lights are in use, the total electricity use for lights is assumed to be halved.

• **Solar gains.** The program makes use of annual solar flux values to work out the usefulness of solar gains. Solar gains are calculated using the equation

$$solar\ gains = FLX * SG * AREA * FRMFC * TXFC * 0.75$$

where

FLX   = solar flux (22 W/m² for north facing windows, 68 W/m² for south facing windows and 40 W/m² for east/west or random orientations)

SG   = fractional shading of the window

AREA   = gross area of the opening

FRMFC = ratio of glazed to total area (frame factors are 0.82 for single-glazed windows, 0.71 for double-glazed windows, 0.50 for single-glazed doors and 0.45 for double-glazed doors)

TXFC   = transmission factor for single and double glazing

0.75   = estimated usefulness of solar gains

continued

- **Sunspaces and conservatories**. The program regards sunspaces and conservatories as additions to the main building. Three effects are considered.

1. Changes to the fabric heat loss caused by an added resistance to the walls and windows covered by the sunspace.

2. Pre-heated ventilation air, providing a mechanism for heat transfer from the conservatory to the house.

3. Heat conducted into the house through the walls and windows covered by the conservatory.

Total extra incidental gains that result from the sunspace are given by the equation

$$CONSGAINS = Q(1-f)(X + (1-X)RC/(RC + RW)) \text{ kWh/day}$$

where

Q = average heating season solar gains into the conservatory, in kWh/day

f = solar gains through covered windows, divided by Q (converted from kWh/day to W)

X = proportion of solar heat transferred into the building by ventilation

RC = total resistance from sunspace to the outside, in °C/W

RW = conductance resistance from the conservatory to the house in °C/W

Solar input and frame factors are estimated on the same basis as for normal solar gains. See above.

The pre-heat fraction is based on a maximum estimate of 60% for south-west orientation (sunspace covering all of the south-west face) and a minimum estimate of 5% for north-east orientation (sunspace covering one-third of the north-east face). The fraction is calculated as the product of the direction factor (DIR) and the proportion of the total elevation (in the direction DIR) that is covered by the conservatory (FRAC). The values used for DIR are

orientation	SW	W or S	NW or SE	N or E	NE
DIR	0.6	0.5	0.4	0.3	0.2

Conservatory resistance RC is calculated as the reciprocal of the sum of all the U A values for all of the elements. Resistance from house to the conservatory is estimated by requiring the operator to input the area of the wall and openings covered.

- **Heating systems and controls**. The program allows the building to be heated by up to two heating systems, one of which must be designated the primary system. The secondary system cannot provide hot water. The primary system is assumed to heat both zones and can also be selected as a hot water heating system. The

heating alternatives cannot accommodate an active solar system. Heat pump systems can be entered with the coefficient of performance replacing the efficiency (it is important to enter the seasonal coefficient of performance and not simply the figure for the device).

The responsivity of heating systems (which affects the demand for heat) is divided into 5 categories.

Gas boiler + rads			Oil systems		
floor mounted	65%	1	boiler + rads	60%	1
wall mounted	70%	1	hot air standard	70%	1*
low energy	70%	1	hot air modulated	70%	1
condensing	85%	1			
Gas hot air			Electric systems		
standard	72%	1*	storage heaters	100%	5
modulated o/p	72%	1	fan assisted	100%	5
			underfloor heating	(calc)	5
Gas room heaters			electric boiler	100%	2
standard	60%	1	ceiling heating	(calc)	4
room sealed	70%	1	room heaters	100%	1
Solid fuel					
auto boiler	70%	2	open fire	35%	3
manual boiler	60%	2	open fire + boiler	60%	3
closed room heater	65%	3	open fire + damper	45%	3

*demand temperature increased by 0.5°C due to poor control

Category 1 in the table above indicates 100% responsivity, 2 indicates 75%, 3 indicates 50%, 4 indicates 25% and 5 indicates zero responsivity. For storage heaters and other off-peak systems the responsivity is assumed to be 25% better if automatic charge controls are fitted.

The controls assigned to a heating system may its efficiency, the demand temperature, or both. The general effects of controls are as follows.

continued

General effects

no room thermostat — boiler efficiency decreased by 5% demand temperature increased by 0.5°C

no programmer — no effect

Control options which alter zone 2 demand temperature

TRVs with no stat — boiler efficiency decreased by 5% zone 2 demand temperature decreased

TRVs with stat + prog — zone 2 demand temperature decreased

zone control — zone 2 demand temperature decreased evening heating reduced by 2 hours

Appliance specific controls

charge control — only on storage heaters, if NOT fitted demand temperature increased by 1°C

damper on open fires — increase fire efficiency to 50%

• **Hot water systems**. For the basic hot water model (right), boiler efficiency is HWEF, distribution efficiency is E2 and tank losses are TL. Solar panels gains are taken from the demand at the tank. Hot water demand HWD is

$$HWD = 1.27N + 2.02 \ kWh/day$$

where N is the number of occupants

The fuel required for providing hot water is then given by

$$HW \ fuel = (TL - SP + HWD)/ (WHWEF * E2)$$

The incidental gains are taken as including 10% of the net hot water demand and are thus given by

$$HW \ gains = HW \ fuel*HWEF*(1-E2) + TL + 0.10*HWD$$
$$= 9TL - SP' + HWD)(1-E2)/E2 + TL + 0.10*HWD$$

Tank losses are estimated using estimates of losses for

insulated and uninsulated tanks to deduce the equation

$$TL = 2.74 + 46.4/(4 + mm) \ kWh/day$$

where mm is the thickness of tank insulation.

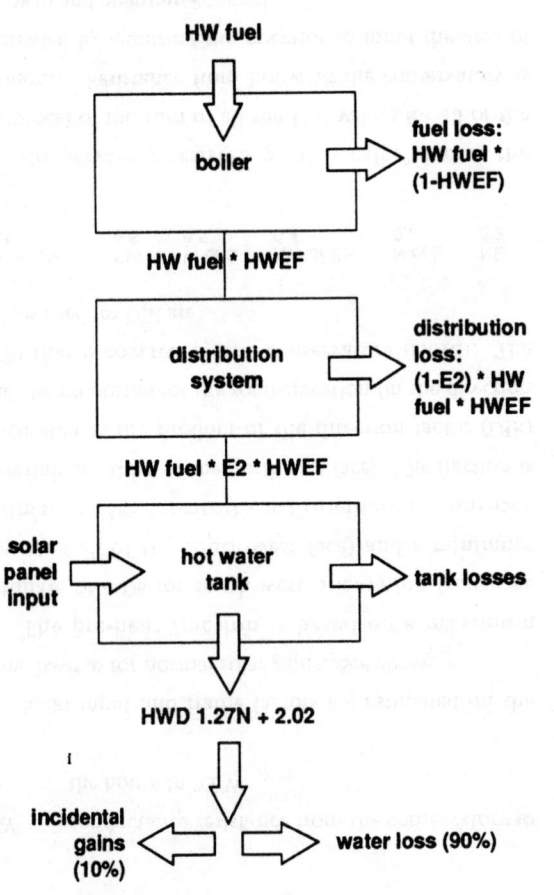

For systems without a hot water tank there are no tank losses, no solar panels and no gains from distribution pipework. The fuel used is simply the hot water demand divided by the instantaneous appliance efficiency. For systems involving a hot water tank the model used for describing hot water systems makes use of a simple energy balance around the tank. Heat is supplied to the tank by a heating device, with its own efficiency, and also by solar panels if present. Heat is lost from the tank by conduction losses and by hot water use. The efficiency of hot water heaters is as follows

immersion heaters — 100%
instantaneous gas — 70%
instantaneous electric — 100%
boilers — same efficiency as for heating

In addition, there is a distribution efficiency associated with tank based systems. This is taken as 95% for normal systems and 85% for thermosyphon systems. This distribution efficiency aims to model the heat losses from the pipes connecting the boiler to the tank; they do not apply for electrically heated tank systems.

The contribution from solar panels has been estimated by an analysis of the performance data on solar panels. The contribution depends on the total panel area and the hot water demand. The larger the demand the greater the contribution per sq. m of panel.

For electric water heating the relative use of on- and off-peak electricity is determined by whether the hot

continued

water tank has one or two immersion heaters, by the size of the hot water tank and by the level of hot water demand (which is in turn determined by the number of people).

• **Energy Cost Index.** The Energy Cost Index (ECI) is calculated from the estimated fuel costs, including standing charges and maintenance costs, per unit floor area. The ECI is defined as

$$ECI = S * (TOT - MR)/TFA$$

where

TOT = total annual fuel costs including standing charges and maintenance costs

TFA = total floor area

S = scaling factor to give an index of 100 for a good low energy house (set equal to 65)

MR = a minimum running cost requirement (for example, £150 to approximate the annual fuel bill for a very small, one person all-electric flat in the United Kingdom)

The expression for TOT is

$$TOT = \Sigma (E_i * V_i) + (F_i \text{ if } E_i > 0) + (m_i \text{ if } i \text{ is primary heating})$$

where

Ei = amount of fuel i used

Vi = fuel cost (£/GJ)

Fi = standing charges for fuel i

Mi = maintenance costs for fuel i

TFA = total floor area

S = scaling factor

To accommodate changes in fuel prices the TOT expression has been indexed to a standard house at 1984 prices: an end-of-terrace, very well insulated house with 100 mm filled cavity walls, 150 mm loft insulation, double glazing and 50 mm underfloor insulation. This standard house is orientated south with minimum over-shadowing and has most glazing on the south face. In 1984, total fuel costs were £365.50. Subtracting the minimum running cost requirement of £150.00 and dividing by the floor area (100 sq.m) gives 2.165 £/sq.m.

This is defined as the standard house with an index of 100 and requires a value of 46 for the scaling factor S. The expression for TOT is multiplied by a weighting function that ensures that if all the fuel costs increase by the same percentage then the ECI of the standard house will remain 100. Since the standard house uses gas as the main from of heating this means that the ECI is fairly independent of price rises in gas and electricity. The full expression for the weighting function is

$$TOT = TOT* \left\{ \frac{366.5}{45.6*PG + 9.1*PE + SG + SE + MG} \right\}$$

where

PG = price of gas

PE = price of electricity

SG = standing charges for gas

SE = standing charges for electricity

MG = maintenance costs for gas

The other expressions of the other indices given in the results section are as follows.

delivered energy = ΣE_i

primary energy = $\Sigma E_i * P_i$

where

P_i = primary energy conversion factors

Primary energy multipliers are taken as the reciprocal of the efficiency of supplying the fuel.

• **Fabric Index.** Instead of a total energy index, the "Fabric Index" (FI) reflects the thermal performance of the house envelope regardless of the appliances and equipment which have been fitted in the house. It therefore includes floor loss and ventilation loss, and also accounts for solar gains and associated factors such as orientation and over-shadowing. The FI can be defined

continued

by the equation

$$FI = 100 \times \frac{\text{space heating requirement as designed}}{\text{space heating if built to Building Regulations}}$$

The space heating requirement is the net useful energy demand: the useful heat demand minus the expected contribution from internal and solar gains. It does not include any appliance or heating system efficiency. To ensure that the FI is independent of the appliances and heating system it is necessary to define the internal gains in a way that does not depend on items such as the water heating system, cooking method etc. The approach adopted in the program is therefore to estimate the total gains on the basis of the floor area.

To estimate the space heating load it is necessary to make assumptions about the responsiveness of the heating system since this affects the background temperature. Responsiveness is assumed to be in category 2 (75%) responsivity) for the as-designed and Building Regulation house. The Building Regulation version is assumed to have the following characteristics.

wall u-value	0.6
roof u-value	0.35
glazing u-value	5.0
floor u-value	calculated for house shape with 0 mm insulation installed
air-change rate	calculated for house as built with no draught-proofing or draught lobby